THE ANTI-ADHD CROSSWORD ACTIVITY BOOK

Crossword for Beginners
with **50** Puzzles

PUZZLE THERAPIST
CROSSWORD | SUDOKU | KIDS & ADULTS

CONTENT

PUZZLE 1

ACROSS

1. Unscramble this word: hsca

5. St. Francis of ___

11. Radar, e.g.: Abbr.

14. Woodstock gear

15. Dirty old men

16. Fill in the blank with this word: "" ___ Haw""

17. Suffer for acting unwisely

19. Suffix with hotel

20. Cry from one who just got the joke

21. Finish this popular saying: "You are what you_____."

22. Finnish architect Alvar ___

24. Subject of some sightings

28. Steinbeck's ___ Small

31. Pabst brand

32. Supreme Court justice nominated by Bush

33. Comedians Bob and Chris

37. Semitic lang.

38. Lymphocyte found in marrow

40. Loud noise

41. Twice-secured

43. "Venice Preserved" dramatist Thomas

44. What ___ surprise!'

45. The English translation for the french word: venger

46. PLASTIC SURGEON

50. Middle name of American architect Frank Wright, born in 1867

51. Object

52. Fill in the blank with this word: ""___ me?""

55. Whitman's "A Backward Glance ___ Travel'd Roads"

56. "Impossible!"

61. Uncle ___

62. Golden Globe winner Pia

63. The English translation for the french word: baluchon

64. Obama's signature health law, for short

65. See 22-Across

66. Verne hero

DOWN

1. Underworld figure

2. Wet nurse

3. What a peeper uses to peep

4. V-J Day pres.

5. Take ___ of absence

6. Three more than quadri-

7. Poli ___

8. Restaurant chain since '58

9. GM: "___ the USA in your Chevrolet"

10. Unscramble this word: seilair

11. The English translation for the french word: rhume

12. Quarterback Rodney

13. Satchel Paige's real name

18. Sun: Prefix

23. Classifier

24. Supplication

25. Ticket taker?

26. You might take it lying down

27. Former Connecticut governor Jodi

28. Tony winner Bert

29. 'Waiting for the Robert ___'

30. Penpoints

34. Out of ___ (away)

35. "King Lear" or "Hamlet": Abbr.

36. Eyelid problem

38. Tiny percentage in the polls

39. Succumb to mind control

42. Coat electrolytically

43. Roundish

45. Recreating

46. The second part missing in the author's name ___ Vargas ___

47. Of the small intestine

48. Pro ___ (perfunctorily)

49. To ___ (perfectly)

52. *I.R.S. form

53. Place for keys and lipstick

54. Safecracker

57. Fill in the blank with this word: ""Win a Date With ___ Hamilton!" (2004 film)"

58. Prefix with center or gram

59. XXX counterpart

60. Univ. research grantor

PUZZLE 2

1	2	3	4		5	6	7	8		9	10	11	12	13
14					15					16				
17				18						19				
20						21		22						
	23				24		25			26	27	28	29	
		30				31		32						
33	34	35			36		37							
38				39						40				
41			42	43					44	45				
46						47		48						
49				50		51				52	53			
		54			55		56					57		
58	59	60	61		62		63							
64					65					66				
67					68					69				

ACROSS

1. Like some muscles

5. Novelist O'Flaherty

9. Capital of ancient Macedonia

14. George Sand's "___ et lui"

15. Toiletries case

16. Take back, in a way

17. ...

19. Up on deck

20. Was two under

21. Gas: Prefix

23. Sowing machine

25. A, in Morse code

30. They're taken in chess

32. Money for money

33. Voltaire's religious view

36. Blessings

38. W. C. Fields film "___ a Gift"

39. Wicked Game' singer Chris

40. Fill in the blank with this word: "___-di-dah"

41. Giant perissodactyls

44. Under ___ pretenses

46. Take turns?

47. Fill in the blank with this word: "___ the side of caution"

49. Hyundai model

51. Unscramble this word: asicol

54. Poet Mandelstam

56. Fill in the blank with this word: ""Sorry, Wrong ___""

58. The Five People You Meet in Heaven' writer Mitch

62. Excoriates

64. Fill in the blank with this word: ""Beer Barrel ___""

65. Environmental sci.

66. Covered with many small figures, in heraldry

67. The English translation for the french word: titan

68. Vodka in a blue bottle

69. They replaced C rations

DOWN

1. Cut down on

2. Stewpots

3. Riviera beach

4. Geared!'

5. More salacious

6. Fill in the blank with this word: ""Am ___ believe Ö?""

7. Fill in the blank with this word: ""___ Lee" (classic song)"

8. Amplified

9. Cheated, in slang

10. U.S.N.A. grad

11. Long-running B'way musical seen by couples?

12. Youth

13. Word of support

18. Stonecrop

22. "The Fountainhead" character

24. Pancreatic enzyme

26. Tag sale site?: Abbr.

27. Headed for ___ (in imminent trouble)

28. Ward and namesakes

29. Inclusive pronoun

31. Room to ___

33. Sad song

34. Fill in the blank with this word: "___ alcohol"

35. George Sand title heroine

37. Twosomes

39. Fill in the blank with this word: "Anti-___ (airport equipment)"

42. Social maven ___ Kempner

43. ___ Empire (land of Suleiman the Magnificent)

44. Unscramble this word: fcuso

45. Widespread African belief

48. Miss America host after Bert Parks

50. News exec Roger

52. Fill in the blank with this word: ""Li'l ___" (Al Capp strip)"

53. Words of assistance

55. Unscramble this word: pkac

57. Some reddish deer

58. Wont

59. What le gendarme enforces

60. Triple-decker, perhaps

61. Volga feeder

63. timid (similar term)

PUZZLE 3

1	2	3	4		5	6	7	8			9	10	11	12
13					14					15	16			
17					18						19			
20				21						22				
23							24				25	26	27	
28				29		30	31		32		33			
		34					35							
36	37	38			39					40				
41			42	43				44	45					
46						47					48	49	50	
51			52		53			54		55				
	56	57				58	59							
60	61			62					63					
64				65					66					
67				68					69					

ACROSS

1. Wanamaker contemporary

5. Temperate

9. Fill in the blank with this word: """___ if you ..." (bumper sticker)"

13. Yeats's land

14. This is either a regional dialect or an odd phrase that can't be understood by looking at the individual parts

16. Manhattan-based fashion co.

17. The Bible's Garden of ___

18. San ___, Calif.

19. Old Chinese money

20. What a mashed potato serving may have?

23. Title bandit in a Verdi opera

24. Torah place marker

25. Youth org. since 1910

28. What's funded by FICA, for short

29. Regulated pollutants, for short

32. Fill in the blank with this word: "___ Desert"

34. Kind of bobsled

36. Look to ___ troublous world': 'Richard

III'

39. Tulsa sch. named for an evangelist

40. Suffixes with ballad and command

41. Churchill subject, with "The"

46. Rummaged (through)

47. Women of Andaluc

48. Pro ___ (for now)

51. The English translation for the french word: ecu

52. Wedge-shaped inlet

54. James of "Gunsmoke"

56. Chowder ingredient

60. Wing, say

62. To-do

63. Fill in the blank with this word: "Et ___"

64. They go by the wayside

65. Theologian Kierkegaard

66. Filmmaker Riefenstahl

67. Supermodel Wek

68. The English translation for the french word: plier

69. Whose woods these ___ think...': Frost

DOWN

1. Varmints, in a classic cartoon line

2. Assistants

3. Richard of TV's "The Real McCoys"

4. Spreader of dirt

5. Bottom of a chest

6. Start to -matic

7. Willingly

8. ___ bag

9. The English translation for the french word: TVHD

10. "Sure"

11. Wichita-to-Omaha dir.

12. Senate Minority Whip Jon

15. Subject for Hume

21. Verb-to-noun suffix

22. Sound

26. Coal-rich German region

27. Landers and others

30. Corp. money managers

31. Some Spanish murals

33. Toothpaste ingredient

34. Set in "Die Fledermaus"

35. Germany's ___ Valley

36. Your majesty'

37. ___ Church, country singer with the #1 hits 'Drink in My Hand' and 'Springsteen'

38. The English translation for the french word: effusif

42. Vortex Mini Ultra Grip Football manufacturer

43. Classicist Hamilton and others

44. City about which Gertrude Stein said 'There is no there there'

45. Org. for the Denver Gold and Chicago Blitz

48. You might give one the slip

49. ___ Chao, only cabinet member to serve through George W. Bush's entire administration

50. Year the first Tour de France was held

53. Belonging to

55. Teatro alla ___

57. Trans-Siberian Railway hub

58. Prefix with scope or logical

59. Jazzy Waters

60. Plaintive cry

61. Where the Azores are: Abbr.

PUZZLE 4

1	2	3	4		5	6	7	8	9		10	11	12	13
14					15						16			
17					18						19			
20				21					22					
		23					24							
25	26	27				28	29				30	31	32	
33					34						35			
36				37					38					
39			40					41						
42		43					44							
	45				46									
47	48			49	50				51	52	53			
54			55				56							
57			58				59							
60			61				62							

ACROSS

1. Beginning on: 2 wds.

5. Winged nuisance

10. The "brains" of 58-Down

14. Seal's opening?

15. Ski jump downslope

16. Schumacher of auto racing

17. Renaissance artist Guido ____

18. Big tournaments for university teams, informally

19. Wait ____!' ('Hold on!')

20. You can't withdraw from them

23. Rite for a newborn Jewish boy

24. Weapon named for its Israeli designer

25. Stately old dance

28. Trust

33. Fill in the blank with this word: ""Charlie ____ Secret" (1935 film)"

34. Two __ for Sister Sara'

35. The Beach Boys' "Barbara ____"

36. King ____

37. Ready to play, you might say

38. Light: Prefix

39. Fill in the blank with this word: "Cambodia's ___ Nol"

40. Tony-winning Rivera

41. Wooer of Olive Oyl

42. They may accept PayPal

44. Constitution writer

45. Pitcher Robb ___

46. Wow

47. Toothpaste sold in Hollywood?

54. Start of the 15th century

55. Realm of Tolkien's Middle-earth

56. Types with fat recording contracts

57. Fill in the blank with this word: "Call ___ evening"

58. Sporty car features

59. Some shoes, for short

60. Saudi monarch

61. Successful job applicant

62. Fill in the blank with this word: "___ Benedict"

DOWN

1. Ornamental cup holder

2. Top-___ (leading)

3. Fill in the blank with this word: "Cosmetics maker ___ Laszlo"

4. First-rate

5. Mouse ___

6. They believed the world was created by Viracocha

7. Noted count, for short

8. Taco stand add-on, in brief

9. Full of spirit?

10. They've got brains

11. Yesteryear

12. Diminutive suffixes

13. U.S. Army E-7

21. Sea eagles

22. Weapons with telescoping bolts

25. Jam

26. Not give ___

27. Woman of letters?

28. Weak ones

29. Zeno of ___

30. Prophet who predicted the destruction of Nineveh

31. Big bill

32. TV newsman David

34. Fill in the blank with this word: "___ Woods National Monument"

37. Victors of 1865

38. Much of kindergarten

40. Lee Van ___ (spaghetti western actor)

41. Tops

43. Under consideration

44. Work on a whaling ship

46. What a cookie cutter cuts

47. Recorded proceedings

48. Potato source

49. Fill in the blank with this word: "___ de boeuf"

50. Title role for Chris Hemsworth

51. Work like ___

52. Cockney greeting

53. Sound of a leak

54. U.K. counterespionage agcy.

PUZZLE 5

1	2	3	4	5		6	7	8	9	10		11	12	13
14						15						16		
17				18								19		
20					21				22		23			
		24	25				26							
	27	28			29	30						31	32	
33				34						35				
36			37		38			39						
40					41		42		43					
44			45	46				47						
	48						49							
50	51					52	53			54	55	56	57	
58			59		60			61						
62			63					64						
65			66					67						

ACROSS

1. Win by ___

6. Weary worker's wish

11. Fill in the blank with this word: ""___, bro?""

14. "The Things We Do for Love" rock group, 1977

15. State sch. in Athens

16. Therefore

17. See 28-Across

19. Grand ___ Opry

20. Vase's handle

21. Fill in the blank with this word: "China's Sun Yat-___"

22. Hospital item

24. Wide receiver Welker

26. The English translation for the french word: parsemer

27. Comedienne joins the picket line?

33. Two qtrs.

34. They greet each other by pressing their noses together

35. Sicilian resort city

36. Fill in the blank with this word: "___

Nurmi, the Flying Finn"

38. Kangaroo ___

39. The ___ near!'

40. Mythical king of the Huns

41. Ltr. accompaniers

43. Taxonomic suffix

44. Takes in recent events

48. In the least

49. Shrink

50. Uses as a target

52. Fill in the blank with this word: "___ pro nobis"

54. Peau de ___ (soft fabric)

58. Fill in the blank with this word: ""___ note to follow ...""

59. Farmer's overalls?

62. W.W. II vessel: Abbr.

63. Sniggled

64. Super Bowl XXV M.V.P. ___ Anderson

65. Fill in the blank with this word: "___ out a win"

66. Huey, Dewey, Louie, Donald and Daisy

67. When liquefied, this form of oxygen with 3 atoms in each molecule is deep blue

DOWN

1. Right back ___!'

2. Liquid ___ (refrigerant)

3. Fill in the blank with this word: "___ probandi"

4. Opposite of fine print?

5. Command level: Abbr.

6. Weapon in the game Clue

7. Fill in the blank with this word: ""Two owls and ___ ...": Lear"

8. Sue Grafton's '___ for Noose'

9. They may be reasonable

10. The Prisoner of Zenda' villain

11. TIME MACHINE ___ SEAT BELT

12. With a 2007 women's water polo title, this California school became the first to win 100 NCAA titles

13. Striking end

18. Direction to an alternative musical passage

23. Fill in the blank with this word: "Avant-garde film maker Maya ___"

25. Zoe's best friend

26. Teller

27. Hit so as to make collapse / Win over

28. to bring or combine together or with something else

29. Most upset

30. Praying figure

31. Sweaters, e.g.

32. Fill in the blank with this word: ""At ___, soldier!""

33. W.W. I plane

37. Vantage points

39. Old laborer

42. Renault 5, in North America

45. Molded

46. Like the Marquis de Sade or the Duke of Earl

47. Unreal

50. Pas ___ (gentle ballet step)

51. Fill in the blank with this word: ""Now ___ you...""

52. Was in the red

53. Funnyman Foxx

55. Fill in the blank with this word: "Catch ___"

56. Pack ___ (quit)

57. Sum, ___, fui

60. Worn out

61. Nanki-___ of 'The Mikado'

PUZZLE 6

1	2	3		4	5	6	7	8		9	10	11	12	13
14				15						16				
17				18						19				
20			21					22						
23							24				25	26	27	
			28		29	30	31		32					
33	34	35		36				37			38			
39			40					41	42					
43					44						45			
46				47	48		49				50			
51					52				53		54	55	56	
			57			58	59	60						
61	62	63			64						65			
66					67						68			
69					70						71			

ACROSS

1. Visual way to communicate: Abbr.

4. Fill in the blank with this word: "Dennis Miller book "___, Therefore I Am""

9. Virtual meeting of a sort

14. Fill in the blank with this word: "___-wolf"

15. Fill in the blank with this word: "Brown-___"

16. Mighty Lak' a Rose' composer

17. Fill in the blank with this word: "Big ___ Conference"

18. Danish astronomer Brahe

19. Annual prize won multiple times by Beyonc

20. The English translation for the french word: vieilli

23. "Candid Camera" co-host Jo Ann

24. Like some victories

28. Like non-oyster months

32. The English translation for the french word: ÈgoÖste

33. Ball catcher

36. Really awful, in rap slang

38. Dismissive exclamation

39. Warplane's cargo

43. Fill in the blank with this word: "___ Beach (D-Day site)"

44. Wee

45. Fill in the blank with this word: "ì___ boom bah!ï"

46. Usual word for an interval when a court suspends business, but doesn't adjourn

49. They may be final or physical

51. Sprightly song

53. Fill in the blank with this word: "___ badge, boy scout's award"

57. Actor Liam's younger kin?

61. The Bee Gees' "How Can You ___ Broken Heart"

64. Worried

65. Verdi's "___ tu"

66. Woolf's "___ of One's Own"

67. Noted Twain portrayer [black]

68. Fill in the blank with this word: "___ el Amarna, Egypt"

69. Lowly ones

70. The ___ Brothers of R & B

71. Fill in the blank with this word: "" ___ I Can Make It on My Own"

(Tammy Wynette #1 hit)"

DOWN

1. Put ___ to (end)

2. Fill in the blank with this word: "Continental ___"

3. U.S. Open champ, 1985-87

4. Incised printing method

5. Wordsworth's 'Rob ___ Grave'

6. Yiddish writer Sholem

7. ___ Bottling Company (Cleveland fixture for over 85 years)

8. Unscramble this word: torpo

9. Sportscaster with the catchphrase "Oh, my!"

10. Condensation

11. The second Mrs. Sinatra

12. French shooting match

13. West ___ Avenue

21. The English translation for the french word: livrÈe

22. Fill in the blank with this word: "Bill ___, TV's Science Guy"

25. They played Ricky Nelson and Bobby Darin

26. Pancreatic hormone

27. Rail supports

29. She, in S

30. Use a knife

31. Yes or no follower

33. Spanish actress Carmen ___

34. Strength of a solution

35. One of the singing Braxton sisters

37. White Sands Natl. Monument state

40. Quote, part 2

41. Put ___ good word for

42. Crossword grid feature

47. Ticks off

48. Sault ___ Marie

50. Kind of I.R.A.

52. Japanese beer brand

54. Scarlett's love

55. Middle of a famous palindrome

56. Weave, in a way

58. Western Indians

59. Fill in the blank with this word: "Dickens heroine ___ Trent"

60. Local theater, slangily

61. No ___!' ('I give!')

62. We'll teach you to drink deep ___ you depart': Hamlet

63. Fill in the blank with this word: "___'wester"

PUZZLE 7

1	2	3	4	5		6	7	8	9		10	11	12	13
14						15					16			
17						18					19			
20					21					22				
			23				24							
25	26	27			28	29	30			31		32	33	34
35				36					37					
38			39					40						
41							42					43		
44					45						46			
			47		48				49	50				
	51	52					53	54				55	56	57
58					59					60				
61					62					63				
64					65					66				

ACROSS

1. With ___ of thousands'

6. Like some nouns: Abbr.

10. Newspaper ad meas.

14. Bad losers

15. Jai ___

16. Soul singer Hendryx

17. Plum, for one

18. Fill in the blank with this word: "___ remover"

19. Stratford-___-Avon

20. Jessica Alba/Chris Evans double feature?

23. Chopping part of a chopper

24. Singer/songwriter Vienna ___

25. Wiener schnitzel meat

28. Fill in the blank with this word: ""___ Have" (Jennifer Lopez #1 hit)"

31. Perfectly good

35. ___ paper (abrasive)

36. Valentine for Val

37. Trepid, dialectally

38. of or relating to sadomasochism

41. Galloping

42. Fill in the blank with this word: "___ 360"

43. Like Brahms's Symphony No. 3

44. Fill in the blank with this word: ""A ___, petal and a thorn" (Emily Dickinson poem)"

45. Minute: Prefix

46. William who wrote "The Dark at the Top of the Stairs"

47. Fill in the blank with this word: "___ Galerie (Manhattan art museum)"

49. World's first carrier with a transpolar route

51. Reach a conclusion by assuming one's conclusion is true

58. TV actress Spelling

59. Luau strings

60. France's ___ R

61. Fill in the blank with this word: ""How sweet ___!""

62. Free ___ (total control)

63. Transmission

64. Slangy denial

65. The English translation for the french word: mÈlodie

66. Cool, very red celestial body

DOWN

1. Left hand's starting position, for a touch typist

2. The English translation for the french word: Cora

3. Make ___ for it

4. of or relating to a septum

5. Something you might get at work

6. Fill in the blank with this word: "___ liquor"

7. Fill in the blank with this word: "Et ___"

8. Holy, in Latin phrases

9. One quoted

10. Curls up

11. Hebrew letter before resh

12. Sinead O'Connor album 'Am ___ Your Girl?'

13. Social maven ___ Kempner

21. Tribal V.I.P.

22. Stuck

25. Travelers' papers

26. Related through the mother

27. This puzzle's theme

29. Sue Grafton's "___ for Lawless"

30. Successor to Clement VIII

32. Pig ___

33. The English translation for the french word: joint torique

34. Facing the pitcher: 2 wds.

36. The English translation for the french word: phylactÈre

37. What Richard III offered "my kingdom" for

39. Like ocher

40. Where Alex Trebek worked as a newscaster

45. Ralph of 'Paths of Glory'

46. Yucca fibers

48. Star Trek' lieutenant

50. Petula Clark's "___ of the Times"

51. Two out of two

52. Dweeb

53. Ruler crowned in 1953, informally

54. West Point rival, for short

55. Fill in the blank with this word: ""I earn that ___": "As You Like It""

56. Wroclaw's river, to Poles

57. Wyo. neighbor

58. Old Teutonic war god

PUZZLE 8

1	2	3	4		5	6	7	8	9		10	11	12
13				14	15						16		
17				18							19		
20							21			22			
		23		24	25		26						
27	28	29		30				31		32			
33				34					35		36		37	38
39			40				41		42				
43				44		45					46		
		47		48		49					50		
51	52				53		54				55		
56					57			58			59	60	61
62				63			64	65					
66				67						68			
69				70						71			

ACROSS

1. "Come Back, Little Sheba" playwright

5. CCCLI tripled

10. Yoko ____

13. Trap, as at a ski lodge

15. Chaplin and others

16. W. Hemisphere assn.

17. "That was the best ice cream soda I ever tasted"

19. Fill in the blank with this word: "1965 #1 hit "____ of Destruction""

20. Focused, at work

21. Trademark forfeited by Bayer under the Treaty of Versailles

23. The English translation for the french word: seine

26. Tuba sound

27. Wharton grad

30. Puts behind bars

32. Basic teaching

33. Repugnant exclamation

34. Former chocolaty Post cereal

36. Fill in the blank with this word: ""____ Stars," #1 hit for Freddy Martin, 1934"

39. With 14-Across, certain crystal

41. Neurological problem

43. Fill in the blank with this word: "Alumni ___: Abbr."

44. Just one little bite

46. Teacher's deg.

47. Year in the reign of Antoninus Pius

49. They get pins and needles

50. Fill in the blank with this word: ""___ Kapital""

51. Italian road

54. Fill in the blank with this word: ""Another ___, Another Show" ("Kiss Me, Kate" song)"

56. Transportation charge

58. Like fresh air

62. The English translation for the french word: kiffer

63. Steely Dan hit, 1980

66. Uno, due, ___

67. Fill in the blank with this word: ""Blame It ___" (Michael Caine film)"

68. Washing machine cycle

69. Suffix with penta-

70. Saint-___ (Faur

71. Fill in the blank with this word: "Boba ___ of "Star Wars""

DOWN

1. Venus de ___

2. Rodenticide name

3. Old Testament book: Abbr.

4. They're flaky

5. Fill in the blank with this word: "Bon ___"

6. Star of "Youngblood," 1986

7. 'Today ___' (California morning show)

8. Fill in the blank with this word: ""___ say it is good to fall": Whitman, "Song of Myself""

9. Sow bug or wood louse

10. Went beyond

11. Girlfriend, in Granada

12. Symbol of generosity

14. Rhinoplasty

18. Place where you might be asked 'Need a lift?'

22. Shooting marble

24. Dolce far ___

25. Roy Wood's band before Wizzard

27. Fill in the blank with this word: "___ Verde National Park"

28. Fill in the blank with this word: ""All ___ are off""

29. Amazes a horror film director?

31. Gets the last of, as gravy

35. Audited

37. Verdi opera

38. Circus reactions

40. Separate the strands of, as rope

42. One enlarging pictures, say

45. Olympic sprinter ___ Boldon

48. Some taters

51. Witch of ___

52. Town near Cape Cod's tip

53. Rocket stage

55. Fill in the blank with this word: "___ but when"

57. Fill in the blank with this word: "Brontî's "Jane ___""

59. Wild goose

60. Y to the max?

61. Site for techies

64. Writer Anais

65. One of the Cyclades

PUZZLE 9

ACROSS

1. Fill in the blank with this word: ""What's in a ___?": Juliet"

5. Discuss, with

9. Stumbled upon

14. When Othello kills himself

15. Je ne sais ___

16. What Spanish athletes go for at the Olympics

17. Fill in the blank with this word: ""___ lay me down...""

18. Fill in the blank with this word: "Albee's "Three ___ Women""

19. Star of "Gigi" and "Lili"

20. Something 7-Down might put down

23. How mini-pizzas are usually cut

24. Peer group?

25. Wagering sites, for short

27. London insurance giant

32. Island near Quemoy

36. Tony Hillerman detective Jim

38. Actress Schneider

39. Clear, as a sky

41. Lure of New Orleans

43. Useful Latin abbr.

44. In ___ of (replacing)

46. Fill in the blank with this word: ""___ a Letter to My Love" (Simone Signoret film)"

47. Post-Taliban Afghan president

49. First name in Egyptian politics

51. Chevrolet model

53. Assistants

58. Contents of guns used in training exercises

63. Words of agreement

64. Fill in the blank with this word: ""Everything will ___" ("Don't worry")"

65. The English translation for the french word: estrade

66. Ointment tube words

67. Spear

68. Deceptively manipulate, with "up"

69. Marshmallow candies in Easter baskets

70. Shook down

71. Rick with the 1976 #1 hit "Disco Duck"

DOWN

1. Fill in the blank with this word: "___-Poo of "The Mikado""

2. Flip ___ (decide by chance)

3. Workweek letters

4. Throw out in the street

5. Sharpness gauge

6. The English translation for the french word: yuan

7. Rapacious

8. Pop music's ___ Vanilli

9. Show disgust

10. Wing-shaped

11. It's the 4-letter term for the thin sheets of dried seaweed in which sushi is wrapped

12. Fill in the blank with this word: "Elvis ___ Presley"

13. Workplace for Reps. and Dems.

21. D-Day time

22. Tipsy

26. Initials in a 1991 financial scandal

28. Yearbook sect.

29. Fill in the blank with this word: "Days of ___"

30. Superhero played by Liam Neeson in a 1990 film

31. Way: Abbr.

32. The English translation for the french word: humble

33. The U.N.'s Kofi ___

Annan

34. Fill in the blank with this word: ""___ she blows!""

35. Furman's partner in brokerage

37. Outback birds

40. Fill in the blank with this word: "___ artery"

42. Five, on a gunslinger's gun

45. Light

48. Diet doctor

50. Acted as an informant

52. Career diplomat Philip

54. Huey, Dewey, Louie, Donald and Daisy

55. Tied, as a French score

56. Tend to again, as an injured joint

57. Exceptional rating

58. Muddles

59. Word before face or heart

60. Fill in the blank with this word: ""Zip-___-Doo-Dah""

61. Informal denial

62. Lacquered metalware

PUZZLE 10

1	2	3	4	5		6	7	8		9	10	11	12	13
14						15				16				
17						18		19						
20						21								
	22				23					24	25	26	27	
28	29				30					31				
32				33				34	35					
36			37				38							
39					40					41				
42					43				44					
45				46					47					
		48	49					50			51	52		
53	54	55					56							
57					58		59							
60					61		62							

ACROSS

1. Fill in the blank with this word: "___ plane"

6. Metric wts.

9. Unscramble this word: rgeae

14. Stimulating

15. Realm of Otto I: Abbr.

16. Oohed and ___

17. Swift as ___

18. Get some quick lodging

20. Fill in the blank with this word: ""___ beaucoup""

21. Where to split hairs?

22. Food preservative: Abbr.

23. This British term for a police spy or informer sounds like it's related to drugs, but it dates from the mid-19th C.

24. Without ___ to stand on

28. Some TV spots, briefly

30. Fill in the blank with this word: "___ 500"

31. Longtime Steelers coach Chuck

32. Motel freebie

33. Cell stuff that fabricates protein, for

short

34. Scratch

36. Fill in the blank with this word: "Cat's ___"

38. Testified

39. Grandfather ___

40. Verdi opera

41. Test in coll., perhaps

42. Writers Fleming and McEwan

43. Made angry

44. British tax

45. IV ___

46. Batting helmet feature

47. The Unsers of Indy

48. Top-of-the-world topper

50. Some soft drinks

53. Star Alliance member

56. Fill in the blank with this word: "___-garde"

57. Stop, in Paris

58. Victory

59. Tried to keep one's seat

60. The English translation for the french word: marÈcage

61. Fill in the blank with this word: "___ pad"

62. The English translation for the french word: tresse

DOWN

1. In ___ (stuck)

2. Walk through water, or measured the ounces of something

3. Russian's neighbor

4. Navratilova and Lendl

5. While no longer linked with Egypt, this country still uses the flag of the United Arab Republic

6. Like the rhythm of a nursery rhyme

7. U.S. invasion site of 1983

8. Based on the number six

9. Tiny battery

10. Member of the legume family

11. The Greek Letter Equivalent : R

12. Want ad inits.

13. Teacher's deg.

19. Female rap trio with the #1 hit "Waterfalls"

25. Part ownership in a bar?

26. Fill in the blank with this word: "___ Glue-All"

27. Clearings

28. Fill in the blank with this word: "___ AC (pharmacy purchase)"

29. Non-vector

30. Latin hymn "Dies ___"

33. Real estate ad abbr.

34. Monster slain by Perseus

35. Therapists' org.

37. Patronage

38. They may be filled with Joy

40. Traffic report source, often

43. Found only in Latin America, they are the largest parrots

44. Witty

46. They dug his grave ___ where he lay': Sir Walter Scott

47. Fill in the blank with this word: ""Dying / Is ___, like anything else": Sylvia Plath"

49. Monsignor's rel.

51. Much may follow it

52. Union Sq. and Times Sq.

53. Texas ___

54. Saccharin discoverer ___ Remsen

55. School days initials

PUZZLE 11

1	2	3		4	5	6		7	8	9	10	11	12	
13				14				15						16
17			18				19							
20							21					22		
23				24	25				26	27				
		28							29			30	31	
32	33	34			35				36					
37			38	39				40						
41						42					43			
44				45				46	47					
	48			49							50	51	52	
53	54			55					56					
57			58	59			60	61						
62						63				64				
	65					66				67				

ACROSS

1. Worldwide workers' grp.

4. The English translation for the french word: flingue

7. Transfer, as a computer file

13. Ugly as ___

14. Underground org.

15. This year's recruits, figuratively

17. Carnegie Hall debut of 1928, with "An"

20. Lack of restraint

21. Tony-winning Hagen

22. Switch ups?

23. [See circled letters]

28. You can see them at marinas

29. Capital of Minorca

32. Some Mercedes-Benzes

35. Zoologist's foot

36. The English translation for the french word: puissant

37. February 4th, to some?

40. Wide-brimmed chapeaux

41. Bunch of sitcom characters

42. See 42-Across

43. When you entered this world: Abbr.

44. Some locks

45. Puts a handle on

48. M

53. What a floozy might show off

55. The English translation for the french word: if commun

56. Opportune

57. Where ETs do knitting and art?

62. What Fred Astaire danced with

63. Ready to relieve 'em of a ___ or two' ('Les Mis√©rables' lyric)

64. Yak in the pulpit?: Abbr.

65. Shelters in snow country

66. Local political div.

67. Le coeur a ___ raisons...': Pascal

DOWN

1. Writer Asimov

2. Transitional state

3. Witching hour follower

4. Old gold coin of Britain

5. Fill in the blank with this word: "___ Theaters (national cinema chain)"

6. Bygone carrier

7. #19 of the Colts

8. The English translation for the french word: pÈnal

9. It

10. Mozart's "L'___ del Cairo"

11. Game sticker?

12. Fill in the blanks with these two words: ""Whatcha ___?""

16. They're not part of the body: Abbr.

18. Actress Schneider

19. Berlioz's 'Les ___ d'ete'

24. Sales slips: Abbr.

25. "Moonstruck" actress

26. "Time's a-wastin'!"

27. "@#$%!," e.g.

30. Fill in the blank with this word: "Catch ___"

31. Wreck-checking org.

32. Is loyal to

33. Peseta : Spain :: ___ : Italy

34. Zigzag activity

36. Withdraws

38. The Bible's Garden of ___

39. Where Loews is "L"

40. In ___ (where found)

42. The English translation for the french word: kiowa

45. Fill in the blank with this word: ""___ a Moon Out Tonight" (1961 hit)"

46. Thoroughly enjoys

47. Use tiny scissors

49. Danish astronomer Brahe

50. Abbr. after Ted Kennedy's name

51. Rep. ___ Hastings of the House intelligence committee

52. Textile workers

53. Gender ___

54. Jai ___

58. Suffix with ether

59. Operations ___ (Army position)

60. Camera setting

61. Wrecker's job

PUZZLE 12

1	2	3	4	■	5	6	7	8	■	9	10	11	12	13
14				■	15				■	16				
17			18						■	19				
20				■	21			22		■	23			
24			■	25	26					■	27			
■	■	28					■	29			■	■	■
30	31	32					■			33	34	35	36
37						■	38	39	40				
41				■	42							
■		43	44	45	■	46				■	■
47	48	49			50				■	51	52	53
54			■	55				■	56	57		
58		59		■	60		61					
62				■	63			■	64			
65				■	66			■	67			

ACROSS

1. Pad ___ (noodle dish)

5. Computer type

9. Lively French dance

14. Fill in the blank with this word: "Bee ___"

15. Fill in the blank with this word: ""I know not why I ___ sad": Shak."

16. Puts up

17. They have whiplike tails

19. Two-time A.L. home run champ Tony

20. Tool attached to a rope

21. Hate

23. Hockey's Krupp

24. Nestle's ___-Caps

25. Flemish : Belgium :: Pashto : ___

28. Pianist Claudio

29. Thoughts on paper?: Abbr.

30. It may keep cafeteria food warm

33. 'Vette option

37. Hostilities

38. Stubborn ___

41. Old record label

42. Gave off

43. TWA rival

46. Pot

47. Site of the War of 1812 Museum

51. Old Dodge

54. Lon ___ of Cambodia

55. Fill in the blank with this word: "___ Lumpur, Malaysia"

56. Others: Sp.

58. Sans ___ (carefree)

60. Golf club used in a bunker

62. Frighten

63. Woman's name suffix

64. Prefix with -itis

65. Attention getters

66. Where Samson wielded a jawbone, in Judges

67. Oxford bottom

DOWN

1. Recipe amts.

2. Toss one's ___ the ring

3. Supreme Court justice nominated by Bush

4. Mae West's '___ Angel'

5. Fill in the blank with this word: ""Knots Landing" actress ___ Park Lincoln"

6. Conglomerations

7. Prepare mentally

8. Put ___ (shove off)

9. Reply in Rome

10. Suffix with hotel

11. Sweep

12. Suburb of Tokyo

13. Germany's University of Duisburg-___

18. Twists

22. 1983 Indy winner Tom

26. Where boys will be boys

27. Mil. V.I.P.

28. Went to Wendy's, say

30. Mao's successor as Chinese Communist leader

31. Treebeard, e.g.

32. Story ___

34. Pres. Hoover's dog King ___

35. Grand ___ Opry

36. Fill in the blank with this word: "___ Xing"

38. Unfading flower

39. Puppeteer Tony

40. In whatever way

42. Use as a model

44. Diet doctor

45. Y.A. Tittle's alma mater

47. Starts to raise, as a hem

48. Nuts

49. Grads

50. With an "I" it's a sweet herb; with an "E" it's a Swiss city

51. Fill in the blank with this word: ""That's an ___!""

52. Sorcerer

53. Val d'___, French ski resort

57. Fill in the blank with this word: "___ penny (very common, in British lingo)"

59. Pride : lion :: clowder : ___

61. Fill in the blank with this word: "Agnus ___"

PUZZLE 13

ACROSS

1. Zachary Taylor, for one

5. Lummoxes

9. Fill in the blank with this word: "___ Bora, wild part of Afghanistan"

13. Winter coating

14. Online shopping center

16. Outdoor shindigs

17. Bears: Lat.

18. First U.S. chief justice

19. Warehouse supply: Abbr.

20. Perfect job for Holden?

23. Weapons prog. since 1983

24. Wintry forecast

25. Mystery writer ___ Jackson Braun

27. Underscore

30. Finish this popular saying: "No man is an_____."

32. ___ Field

33. Fill in the blank with this word: "___-Cola"

35. Bed cover

38. Right in every detail

41. Who wrote "In the country of the blind the one-eyed man is king"

43. Sharp-___

44. Unexciting

46. Fill in the blank with this word: ""Chances ___," 1957 #1 hit"

47. Golf's Sorenstam

49. Fill in the blank with this word: "___ honorable (formal apology)"

52. Civics, e.g.

54. Fill in the blank with this word: "___ basin"

56. Sounds of doubt

57. Maternity ward arrival

62. Seal's opening?

64. Fill in the blank with this word: "___ living"

65. Village Voice theater award

66. Waterfall sound

67. Jordanian tongue

68. Fill in the blank with this word: "___ of the state"

69. Van ___, Calif.

70. Verb-to-noun suffix

71. W.W. II city on the Vire

DOWN

1. Trounce

2. Prefix with scope or logical

3. Fill in the blank with this word: ""Now ___ you...""

4. The Dolphins retired his #12

5. Oscar winner of 1990

6. Whack-___

7. Zealous devotee

8. Wow

9. Female rap trio with the #1 hit "Waterfalls"

10. Gardener, say

11. Joe ___, ex-Royals third baseman known as the Joker

12. Words after ugly or guilty

15. Geology pioneer Sir Charles ___

21. Went off

22. Rope with a loop

26. Keglers' places

27. Wallop

28. Well-___

29. President's beam?

31. Writer's supply: Abbr.

34. Ural River city

36. Western scene painter Peter

37. Go on a vacation tour

39. Front

40. Oilman Kashoggi

42. Motocross, e.g.

45. Popular breakfast food

48. Fill in the blank with this word: ""___ little silhouetto of a man" ("Bohemian Rhapsody" lyric)"

50. Finish this popular saying: "Time and tide wait for no_____."

51. You can lean on them

52. Two-time Tony winner George

53. State sch. in Athens

55. The English translation for the french word: ionique

58. Tech expert, as it were

59. Like ___ out of hell

60. Practice logrolling

61. Tokyo, once

63. Fill in the blank with this word: ""48___""

PUZZLE 14

ACROSS

1. One of a series of joint Soviet/U.S. space satellites

5. Some TV spots, briefly

9. Germany's ___ Valley

13. Wrongfully take

14. Rubaiyat' rhyme scheme

15. Wash up against

16. Start of an Oscar Wilde quote

19. Dutch city

20. Unscramble this word: edra

21. Quark/antiquark particles

22. Unctuous

23. Six-stringed instrument

24. Airheads

27. Liza ... with a Z

31. Senator of Watergate fame

32. Tutsi foe

33. They dug his grave ___ where he lay': Sir Walter Scott

34. Scold severely

38. The Tar Heels: Abbr.

39. Wood sorrels

40. John's neighbor

41. Mexican dance music

44. Salutation abbreviation

45. What a D.J. speaks into

46. Italian river valley in W.W. II fighting

47. Fill in the blank with this word: "___ Lupin, the Prince of Thieves"

50. Wise one

51. Works for an ed.

54. Jury's reply to the judge

57. Fill in the blank with this word: "___-wip"

58. Fill in the blank with this word: "___ moss"

59. Jewish youth org.

60. Gray ___

61. WWW addresses

62. Chimney-top nester

DOWN

1. Four after the first?

2. Crocus or freesia, e.g.

3. Fill in the blank with this word: "Auvers-sur-___, last home of Vincent van Gogh"

4. When the Roman general Drusus Senior died

5. Triptych trio

6. Skewered Asian fare

7. Under the covers

8. With 100-Across, Naples opera house Teatro di ___

9. Tangle (with), in the country

10. What have you been ___?'

11. Chemistry Nobelist Otto

12. Ways: Abbr.

15. Use, as a cot

17. Fill in the blank with this word: "___ Lines (big name in cruises)"

18. Countdown term

22. Harriet's mate

23. Saint of dancers

24. Free of sticky stuff

25. Fill in the blank with this word: "___ Spalko, Indiana Jones villainess"

26. Dual-purpose family room device

27. Home of the National Stock Exchange of India

28. Yorkshire city

29. "American Beauty" hero

30. Sci. facilities

32. Vaughn's 'Psycho' co-star

35. Rink game

36. This parish of Louisiana shares its name with a European peninsula

37. Fill in the blank with this word: "___ prius (trial court)"

42. Staunton of 'Vera Drake'

43. ___ a stinker?' (Bugs Bunny catchphrase)

44. Powers

46. Union part

47. You'll want to munch on petha & gazak, signature sweets of this Taj Mahal city

48. Sorry soul

49. Yeas or nays

50. Fill in the blank with this word: "Battle of the ___, opened on 10/16/1914"

51. Those running the place: Abbr.

52. Fill in the blank with this word: ""___ gut" (German praise)"

53. Wyo. neighbor

55. Kwik-E-Mart clerk on "The Simpsons"

56. Trains: Abbr.

PUZZLE 15

1	2	3	4		5	6	7	8			9	10	11	12
13					14					15		16			
17				18							19			
20				21					22					
			23					24						
25	26	27			28	29						30	31	32
33					34						35			
36				37						38				
39			40					41					
42		43					44						
		45					46						
47	48	49				50	51				52	53	54
55					56				57				
58				59					60				
61				62					63				

ACROSS

1. Periodic table no.

5. World War II weapon

9. Refill when you don't really need to

13. Trucks in reverse do this to signal people behind them; the Rostra obstacle sensing system does it to alert the driver

14. Turkish V.I.P. of old

16. Types with fat recording contracts

17. 62060

20. Barges

21. Web-footed rodent

22. Sycophant

23. Finish this popular saying: "You are what you_____."

24. Fill in the blank with this word: "Big ___"

25. What 17-, 24-, 48- and 57-Across are

33. Spanish direction

34. Warning sensations, in medicine

35. Wiretapping grp.

36. Had an unquiet sleep

37. Noted Detroit brewer

38. The English translation for the french word: maille

39. Smallest NATO member by population

40. The Jungle Book' bear

41. Preserve, as preserves

42. Whence the song "Mack the Knife," with "The"

45. Au ___

46. Grand ___ Opry

47. Words finger-drawn on a dirty car

50. Endangered Asian deer

52. When it's broken, that's good

55. "What?!"

58. What has made some people miss the mark?

59. Weave, in a way

60. Unscramble this word: etdb

61. The Syr Darya, one of central Asia's major rivers, flows through 3 countries whose names end in this syllable

62. Women of Andaluc

63. Flightless bird: Var.

DOWN

1. Some military defenses, for short

2. Sound system brand

3. Walks down the aisle

4. Object

5. Wears jauntily

6. Lake ___, head of the Blue Nile

7. "Scratch that!"

8. Whaler's org.

9. Tone deafness

10. Whiff

11. Graceful bend

12. The Beatles' "Back in the ___"

15. Palestinian nationalist group

18. Metalworking tool

19. "Quite possibly"

23. Theater's ___'acte

24. The name of this dark brownish red comes from the French word for "chestnut"

25. Fill in the blank with this word: ""I'll have ___ this one out""

26. Fill in the blank with this word: "1950 Medicine Nobelist Philip Showalter ___"

27. 'God ___ refuge ...': Psalms

28. British society magazine

29. Second-largest lake in North America

30. Short online message

31. Hit ___ note

32. Phi Beta ___

37. Sound

38. Rapper ___ Fiasco

40. See 71-Across

41. This word refers to extreme opposites or a type of bear

43. Writer and director of "Julie & Julia," 2009

44. Unsophisticates

47. Things that open and close yearly?

48. When Parisians typically vacate

49. Quran chapter

50. Fill in the blank with this word: "___-fry"

51. Venezuela's ___ Margarita

52. Work with feet

53. Seine feeder

54. V preceder

56. Cries of pain

57. Suffix with sulf-

PUZZLE 16

ACROSS

1. Not one's best effort, in coachspeak

6. Half a ring

9. Std. on food labels

14. Unscramble this word: ranbu

15. Hall-of-Fame basketball coach Hank

16. The treaty of this city ending the War of 1812 returned all lands captured by Britain & the U.S. during the war

17. Veggie sub request

18. Swiss stream

19. Fill in the blank with this word: "Biff

___, Arthur Miller character"

20. Pick from a deck

23. Some valuable 1920s-'40s baseball cards

24. Skunk feature

25. Red ___

28. Worrying sound to a balloonist

29. Uniform: Prefix

30. City where A.A. was founded

32. Fill in the blank with this word: ""___ could have told you that!""

34. Vitamin a.k.a. riboflavin

35. See 20-Across

41. Take up and hold, chemically

42. Fill in the blank with this word: ""The Rights ___""

43. Unwed fathers

47. You're killing me,' textually

48. Hoover ___

51. Fill in the blank with this word: "Amniotic ___"

52. Worries

54. Fill in the blank with this word: ""And ___ word from our sponsor""

55. Campus group in its typical state?

58. Jazz's ___ Lateef

60. Fill in the blank with this word: "___ Onassis, Jackie Kennedy's #2"

61. Fill in the blank with this word: ""That's what ___!""

62. Two-___ (court situation)

63. Racecar driver ___ Fabi

64. Ward off

65. Fill in the blank with this word: "Caput ___ syndrome (arm problem)"

66. Wine: Prefix

67. Wrinkle removers

DOWN

1. Scams

2. Old English silver coins

3. Cancels

4. Yum-Yum, Peep-Bo and Pitti-Sing in "The Mikado"

5. Requests for developers: Abbr.

6. Sound from a receiver

7. Hornet, e.g.

8. Ice cream flavor Cherry ___

9. Not so attractive

10. Fill in the blank with this word: "___-in"

11. Scraps

12. Satellite ___

13. When some stores open

21. Germany's University of Duisburg-___

22. Tollway: Abbr.

26. Fill in the blank with this word: ""___ lay me down...""

27. Fund-raising suffix

29. Fill in the blank with this word: "___ Toguri (Tokyo Rose)"

31. Scottish seaport known for its single-malt Scotch

32. Wittenberg's river

33. 37-Across's age on May 29, 2003

35. Sugar amts.

36. Fill in the blank with this word: ""Que ___ es?" (Spanish 101 question)"

37. John ___, designer of the warship Monitor

38. [See title, and proceed]

39. Wax theatrical

40. Fill in the blank with this word: "___-Mart"

44. Fixed charge

45. Fill in the blank with this word: "___ particle"

46. Prefix with sphere

48. Triangular chip

49. Stir

50. Mums

53. Silk dress: Var.

54. N.Y. Jet or Phila. Eagle

56. Insurance giant

57. What ___ tell you?'

58. Xavier Cugat film "___ Were Never Lovelier"

59. Having no bounds: Abbr.

PUZZLE 17

ACROSS

1. W.W. II city on the Vire

5. Reno and Kennedy, e.g.: Abbr.

8. Unlocks, poetically

12. toward the mouth or oral region

13. Tore into

15. With 69-Across, 1930s-'50s bandleader

16. Apartment building feature

18. The rain in Spain

19. Lassie, for one

20. Un-PC behavior

22. Taken ___

23. Depend upon, as a decision

25. Trifling

27. Things gained and lost in football

28. Paderewski's "Minuet ___"

30. Weaken

31. Some wild parties

33. They may be hard to change

34. Uncle ___

37. Question from the picked-upon

39. Fill in the blank with this word: ""Humanum ___ errare""

40. Tailors

43. Winner of 2.7% of the 2000 presidential vote

46. Fill in the blank with this word: "___-robe (Calais closet)"

47. Whiz-bang

48. Talk follower

52. Strategic Iraqi port just across the Kuwaiti border

54. Lizard

56. Fill in the blank with this word: "___ system"

57. Large Indonesian island

59. Year in Edward the Elder's reign

60. Letter before shin

62. Quarterback's ideal

64. Purcell's "___ and Welcome Songs"

65. Words to the maestro

66. Unusual shoe spec

67. ___ Institute (astronomers' org.)

68. W. C. Fields film "___ a Gift"

69. Misses

DOWN

1. Piano

2. ___ night (bar attraction)

3. Storage spot

4. Fill in the blank with this word: ""___ to Billie Joe" (1967 #1 hit)"

5. Sixth Jewish month

6. with the mouth wide open as in wonder or awe

7. Trial figure

8. Fill in the blank with this word: "Costa Rica's ___ Peninsula"

9. Add numbers to, say

10. Copies

11. Least of all

13. Operator's accessory

14. Anaconda loser, almost certainly

17. Tool for some group mailings

21. Fill in the blank with this word: "*Ace ___ Stories (old detective pulp magazine)"

24. Writing: Abbr.

26. Writer Chinua Achebe, by birth

29. *Group with the 2000 #1 hit "It's Gonna Be Me"

32. Utterances around baby pictures

33. Like associates, on some e-mails

34. They first bloom when 50-75 years old

35. Eliot novel

36. Squirrellike monkey

38. Toscanini and others

41. Stat

42. Some are green

44. Unscramble this word: elqua

45. Soldier

47. Sure application spot

49. Site of two ecumenical councils

50. Woman in distress?

51. Members of the carrot family

53. The English translation for the french word: sushi

55. Trapped like ___

58. River isles

61. Fu-___ (legendary Chinese sage)

63. The English translation for the french word: simbleau

PUZZLE 18

ACROSS

1. Fill in the blank with this word: "___ nothing"

6. Six-stringed instrument

10. The English translation for the french word: imam

14. Yellow's opposite

15. Old laborer

16. This small rodent whose name rhymes with mole is closely related to the lemming

17. HO

20. One succumbing to 6-Down

21. Puerto ___

22. Two semesters

23. Just ___ should be

25. Sodium hydroxide, to chemists

27. Sherlock Holmes phrase, when on a case

33. Rodeo performer

34. Yawn

35. Environmental sci.

37. Winding road shape

38. Oh, really?!'

42. Tony-winning Hagen

43. Western Indian

45. Tiny ___

46. First black N.F.L. Hall-of-Famer ___ Tunnell

48. Declaration from a die-hard beer drinker?

52. Monk's hood

53. Who's there?' answer

54. Provide with a new soundtrack

57. Must've been something ___'

59. Tavern order

63. 52-Across, starting 1/20/09

66. Dos minus dos

67. I'll come to you ___': Macbeth

68. Katey of "Married

69. Pop music's Bee ___

70. Ukr. and Lat., once

71. Over here...'

DOWN

1. Some military defenses, for short

2. Scientology's ___ Hubbard

3. Work at the docks

4. Surplus

5. Theologian's subj.

6. Cin

7. Biblical patriarch whose name means "he will laugh"

8. Small planted bulb

9. Fill in the blank with this word: "Basketball's ___ Elmore"

10. Son of Cedric the Saxon

11. The English translation for the french word: moue

12. The English translation for the french word: algue

13. Physics Nobelist Simon van der ___

18. Well-groomed

19. This can be a careless mistake or a foolish person who might make one

24. Wrapped garment

26. Spa sounds

27. Smartphone introduced in 2002

28. Popular landscaping plant

29. Fill in the blank with this word: "England's ___ Downs"

30. Mosque priests

31. Eyes

32. On a scale of one ___

36. Singer k. d. ___

39. Beginning on: 2 wds.

40. Gobs and gobs

41. 1936 Olympics hero

44. Portuguese money replaced by euros

47. Lure of New Orleans

49. Soliloquy starter

50. They were brought down by Olympians

51. Up and ___!'

54. Cockney greeting

55. Unusual shoe spec

56. Truth or ___ (slumber party game)

58. Get an ___ effort

60. Fixes, in a way

61. Wood sorrels

62. Who is John ___?' (question in 'Atlas Shrugged')

64. Fill in the blank with this word: "___ Tafari (Haile Selassie)"

65. Walken's gift in "The Dead Zone"

PUZZLE 19

1	2	3	4		5	6	7	8			9	10	11	12
13					14					15				
16			17					18						
19				20				21						
		22					23							
24	25	26				27				28	29	30		
31					32					33				
34				35				36	37					
38				39				40						
41			42	43			44							
		45				46								
47	48	49				50				51	52	53		
54				55				56						
57				58				59						
60				61				62						

ACROSS

1. You can get a grip on it

5. Fill in the blank with this word: "Call ___ evening"

9. Waffle House alternative

13. Prefix with scope or logical

14. What's expected

15. Unicellular organism

16. Song from "Anything Goes"

19. Fill in the blank with this word: ""___ Kapital""

20. Vigoda and Fortas

21. Word with aunt or voyage

22. Wow

23. The English translation for the french word: Serbe

24. Good watchdog

27. Make a comeback

31. Cornice support

32. Young salmon

33. Visual way to communicate: Abbr.

34. 1979 George Hamilton film

38. Kenyan president Daniel arap ___

39. Wagner soprano

40. "D

41. Words after "The end"

44. "Peer Gynt" dancer

45. Workplace for Reps. and Dems.

46. Whisked mixture

47. Work on more, as a farrier might

50. Whack

51. Texas ___

54. "Blue" boxer

57. Senate house in ancient Rome

58. Fill in the blank with this word: ""To ___ not to ...""

59. This shepherd buys the farm in Genesis 4:8

60. Gulf of ___, body of water next to Viet Nam

61. In-box fill: Abbr.

62. Fill in the blank with this word: ""___ known then...""

DOWN

1. The Kennedys, e.g.

2. Vincent Lopez's theme song

3. Top prizes at the Juegos Ol

4. What a violinist may take on stage, in two different senses

5. Where many shots are taken

6. Maguire of 'The Great Gatsby'

7. Some airport data: Abbr.

8. Name placeholder in govt. records

9. Tipple

10. Retained

11. Poulenc's "Sonata for ___ and Piano"

12. Stooge

15. The English translation for the french word: stupeur

17. PbS

18. Foreign dignitaries

22. Word with strings or horns

23. Missionary Junipero ___

24. Former Swedish P.M. Olof ___

25. Pry

26. Early third-century year

27. Worker's demand

28. John ___, Doris Day's co-star in "The Pajama Game"

29. The Sun, for example

30. Fill in the blank with this word: ""Maria ___," Jimmy Dorsey #1 hit"

32. "Candid Camera" co-host Jo Ann

35. Fill in the blank with this word: "19th-century Swedish writer Esaias ___"

36. Held to its full time, in music

37. Vitamin whose name sounds like a bingo call

42. Toast to one's health

43. Alley Oop's mate

44. TechCrunch and Huffington Post readers, presumably

46. Register anew

47. Hi-fi spinners: Abbr.

48. Prefix with lateral

49. Quran chapter

50. Tournament passes

51. Rubaiyat' rhyme scheme

52. Not worth ___ cent

53. Year the oldest college in the Americas was founded, in Mexico City

55. The English translation for the french word: guichet automatique bancaire

56. Yup's alternative

PUZZLE 20

ACROSS

1. Reindeer herder

5. The ____ Love' (R.E.M. hit)

9. Well-groomed

13. Greek war goddess

14. Fill in the blank with this word: "____-robe (Calais closet)"

16. Obtained: Fr.

17. Raids at the Colossus?

20. Fill in the blank with this word: "____ TomÈ"

21. Fill in the blank with this word: ""____

the lookout!""

22. Visited by the vice squad

23. Keats's "Ode on a Grecian ____"

24. Wine holder

25. Theater question

33. Horror-film character

34. Film score composer Morricone

35. Worldwide workers' grp.

36. Vest feature

37. Red Bordeaux

38. Use tiny scissors

39. Victory sign

40. Wake Up Little ___' (#1 Everly Brothers hit)

41. Fill in the blank with this word: "___ good example (shows the proper way)"

42. Tasteful bedclothes?

45. The "H" in "M*A*S*H": Abbr.

46. Fill in the blank with this word: "___ Cayes, Haiti"

47. Comparable to a cucumber

50. Take ___!' (track coach's shout)

52. Fill in the blank with this word: "___ Khan"

55. Bird on the links?

58. Fill in the blank with this word: ""___ Lee" (classic song)"

59. Deletions

60. That: Lat.

61. Latin grammar task: Abbr.

62. Fill in the blank with this word: "___-majestÈ"

63. Dilbert co-worker

DOWN

1. Visitors learn how to make these floral necklaces at Senator Fong's plantation & gardens

2. Trollope's "Lady ___"

3. Nickname for JosÈ

4. Nanki-___ of 'The Mikado'

5. Nash and others

6. Ovid's family name

7. The Isle of Man's Port ___

8. Insurance card info

9. This Russian term for a sled drawn by 3 horses has made its way into English

10. Funnyman Foxx

11. Summer cooler

12. Yellow squirt?

15. Unpredictable

18. Month between marzo and mayo

19. Word repeated before some relatives' names

23. Slangy denial

24. Went by dugout

25. Fill in the blank with this word: "1997 Aaliyah hit "The One ___ My Heart To""

26. Wail on a 33-Down

27. Rourke's co-star in 'The Wrestler'

28. Escorts to a second-floor apartment, say

29. No ___ sight

30. Kunta ___ of "Roots"

31. Fill in the blank with this word: ""I Still See ___" ("Paint Your Wagon" tune)"

32. Without face value

37. Kellogg's cereal

38. Wraps (up)

40. Single-master

41. -

43. Infamous 1984 gas leak site

44. Slip by

47. Nav. ___

48. Oscar-nominated actress for "Leaving Las Vegas"

49. Publisher's concern: Abbr.

50. Tropical fever

51. W.W. II vessels

52. One going to the dogs?

53. Major-___ (bigwig)

54. Optimistic

56. Lon ___ of Cambodia

57. Shaggy Scandinavian rug

PUZZLE 21

ACROSS

1. The English translation for the french word: tige

6. Group formed at C.C.N.Y. in 1910

9. Comic Smirnoff

14. Vocal qualities

15. Old New Yorker cartoonist Gardner ___

16. Fill in the blank with this word: "___ vincit amor"

17. Trim to fit, perhaps

18. Horizontally: Abbr.

19. Fill in the blank with this word: ""Whither thou ___...""

20. Rich man's wife, often

23. Fill in the blank with this word: ""Take ___ a sign""

24. Tracy's "Tortilla Flat" co-star

25. Worker in a garden

28. Napoleonic marshal

29. Fill in the blank with this word: "___-de-sac"

30. Ballpark vendors' offerings

32. Fill in the blank with this word: "Caput ___ syndrome (arm problem)"

34. Scottish singletons

35. Sign at a roach motel?

41. Fill in the blank with this word: """This guy walks into ___ "

42. Some processor chips

43. Santa ___ (Baja California port)

47. Word repeated in Emily Dickinson's "___ so much joy! ___ so much joy!"

48. Org. for mom-and-pop stores

51. Stowe girl

52. Take the top off, in a way

54. Rousing cheers

55. Boatswains, e.g.

58. Brave one

60. Australian singer Christine

61. a virus that is parasitic in bacteria

62. Wrinkly snack

63. TV's "Emerald Point ___ "

64. Trickles

65. Showed again

66. Wind dir.

67. Saint-___ (Faur

DOWN

1. Yalta Conference attendee

2. Yet

3. Tomorrow

4. Diarist Samuel

5. Fill in the blank with this word: "___ perpetua (Idaho's motto)"

6. Chorus parts

7. X

8. The English translation for the french word: paria

9. Kind of smoothie

10. Omnia vincit ___

11. Patella

12. Sue Grafton's '___ for Outlaw'

13. Winery sight

21. Threw

22. With 17-Down, a temporary urban home

26. Elbe tributary

27. Sum, ___, fui

29. Gulager of TV's "The Virginian"

31. Three of these make an O

32. Shadow

33. You may be asked to arrive 90 mins. prior to this

35. The item seen here, or the taste it might have if made with lemons

36. Wanderers

37. Convenient, in a way

38. Defunct

39. Bribed

40. Some Mercedes-Benzes

44. Writer of "The 95 Theses"

45. Fill in the blank with this word: "1099-___ (tax form sent by a bank)"

46. These people who settled India c. 1500 B.C. were taken for a superior race in Nazi pseudoscience

48. Vulgarity

49. She won five Emmys for her sitcom title role

50. Weigh

53. Chaplin and others

54. United States biochemist (born in Spain) who studied the biological synthesis of nucleic acids (born in 1905)

56. Trix alternative?

57. NYSE banner events

58. Orderly supervisor, maybe: Abbr.

59. Saccharin discoverer ___ Remsen

PUZZLE 22

1	2	3	4			5	6	7		8	9	10	11
12					13	14				15			
16				17						18			
19				20				21		22			
23			24				25		26				
27					28	29			30				
		31		32				33					
34	35	36		37		38				39			
40			41	42		43		44	45				
46					47			48		49	50	51	
52				53		54	55						
56				57		58				59			
60		61		62				63					
64				65				66					
67				68				69					

ACROSS

1. Fill in the blank with this word: "Arctic ___"

5. Holy _____ : Bovine

8. Ancient city with remains near Aleppo

12. Now ___ me down to sleep'

13. Think out loud

15. Where you might be among Hmong

16. Not-so-super bowl?

18. London's ___ of Court

19. Fill in the blank with this word: "BBC : Britain :: ___ : Italy"

20. The English translation for the french word: coquerie

22. Fill in the blank with this word: "Fiddle-de-___"

23. Without money changing hands

25. Words of contentment

27. You can take them in stride

28. Fill in the blank with this word: "Basketball's ___ Elmore"

30. Old newspaper sections

31. Oarlock

33. Whom People magazine once named

the world

34. Withdraw

37. One way to break ties

39. Le coeur a ___ raisons...': Pascal

40. Tubular food

43. Fill in the blank with this word: ""This foolishness must ___ once!""

46. Fill in the blank with this word: "2003 Nick Lachey hit "___ Swear""

47. Prefix with center or gram

48. String bean's opposite

52. Novelist S

54. Skateboarding tricks

56. Year that Augustus exiled Ovid

57. Take the blame for

59. Fill in the blank with this word: "Actress ___ Ling of "The Crow""

60. Personal air

62. Impossible to change

64. Fill in the blank with this word: ""Just you wait, ___ 'iggins...""

65. This cavalry weapon was inspired by the Turkish scimitar

66. Fill in the blank with this word: ""To ___ not to ...""

67. Univ. paper

68. Y. A. Tittle scores

69. Wearers of four stars: Abbr.

DOWN

1. River past the ruins of Nineveh

2. Chemist's extracting solvent

3. Rhea, e.g.

4. WSJ competitor

5. This farewell word first appeared in an English text in Hemingway's "A Farewell to Arms"

6. To be returned

7. W can be a vowel in it

8. Ransom ___ Olds

9. South-of-the-border outlaws

10. Wanting company

11. Deem

13. toward the mouth or oral region

14. "Goodnight, Irene" singer, briefly

17. What online shoppers may spend

21. Zaragoza's river

24. Sch. assignment

26. Myrna of "The Thin Man"

29. Sniggled

32. Fill in the blank with this word: "___-Wan Kenobi"

33. Wall Street earnings abbr.

34. Well-regarded

35. Native of one of the Gulf States

36. Horse focuser

38. Theoretically

41. Wall Street org.

42. The English translation for the french word: miso

44. Naturals

45. Mai ___

47. The Secret Agent' author

49. Like some folders

50. Zigzag, in a way

51. Willows

53. The English translation for the french word: vrille

55. Sein : German :: ___ : French

58. High-rise locales

61. Yankee insignias

63. Hall-of-Fame basketball coach Hank

PUZZLE 23

1	2	3	4	5		6	7	8	9	10		11	12	13
14						15						16		
17				18								19		
20							21		22					
		23		24	25			26						
27	28	29		30					31					
32			33		34		35			36		37	38	
39				40			41		42					
43				44		45			46					
		47	48			49		50			51			
52	53					54					55			
56				57				58			59	60	61	
62				63		64	65							
66				67						68				
69				70						71				

ACROSS

1. Thin iPods

6. Fill in the blank with this word: "___-Kettering Institute"

11. Fill in the blank with this word: "___-wolf"

14. Fill in the blank with this word: ""___ Wanna Cry," 1991 #1 song"

15. Spanish direction

16. Texas ___

17. Chamber of commerce policies

19. End of many an E-mail address

20. Springsteen's E ___ Band

21. Lover of beauty

23. School skipper

26. Fill in the blank with this word: "___-guided"

27. Southern ___

30. Within: Prefix

31. Brine

32. Titled Turks

34. Natl. negotiator

36. The Third of May 1808' painter

39. "Survivor" setting, 2004

41. Humorist Ward

43. Son of Noah

44. Caesar's empire, for short

46. Union requirement, maybe?

47. Fill in the blank with this word: "Betty ___"

49. The Beatles' "Back in the ___"

51. Sue Grafton's '___ for Noose'

52. It's hot in Paris

54. Elegantly groomed

56. Shakespeare's "The Rape of ___"

58. Vegetarian's stipulation

62. Litmus bluer: Abbr.

63. Citrine

66. Reply facilitator: Abbr.

67. Vampire ___ (fanged fish)

68. Skin problem

69. Ukraine, e.g., formerly: Abbr.

70. Univac's predecessor

71. More cunning

DOWN

1. Penpoints

2. Poke-___!' (kids' book series)

3. Koh-i-___ diamond

4. Word go

5. Noted Carmelite mystic

6. Young women's grp

7. Wreath

8. Native of the central Caucasus

9. They're often pressed for cash

10. Swiss multinational

11. Some clerks

12. When you "make" this, you go with speed

13. OPEC V.I.P.

18. Vingt-___ (blackjack)

22. Fill in the blank with this word: "___-Dazs"

24. Total

25. There's ___ in team'

27. Quicken Loans Arena cagers

28. Wet nurse

29. Red-dogger, e.g.

31. Type of 35mm camera

33. London Sun tidbit

35. Zoroastrian

37. Doctor Zhivago

38. Wriggly biters

40. The English translation for the french word: c.‡.c.

42. They're good at hits

45. Status ___

48. Von Rothbart turned her into a swan

50. Fill in the blank with this word: "___ even keel: 2 wds."

52. Teacher's charges

53. Sinuous dances

54. Three more than quadri-

55. Yahoo! or AOL offering

57. Filmmakers: Joel & Ethan

59. About

60. To ___ (unerringly)

61. Russian ruler: Var.

64. The Era of ___ (1964-74 Notre Dame football)

65. Fill in the blank with this word: "Country music's ___ Brown Band"

PUZZLE 24

1	2	3	4		5	6	7	8	9		10	11	12	13
14					15						16			
17				18						19				
20										21				
22				23				24					25	26
27			28			29	30	31				32		
			33	34	35		36				37			
	38	39				40								
41									42					
43				44			45				46	47	48	49
50			51					52	53	54		55		
		56				57	58				59			
60	61				62									
63					64						65			
66					67						68			

ACROSS

1. Fill in the blank with this word: "___ David"

5. Wood-shaping tools

10. Some hosp. records

14. When expected

15. Watch word

16. Sodium hydroxide, to chemists

17. ORDINAL

20. They have extensions

21. Sue Grafton's "___ for Alibi"

22. Spanish queen until 1931

23. Japanese prime minister Taro ___

24. Morning ___ (bathing, combing the hair, etc.)

27. People: Prefix

29. Favored, with "on"

32. Fill in the blank with this word: "___ 1 (Me.-to-Fla. highway)"

33. Mozart's "L'___ del Cairo"

36. Develop sores

38. Mule, e.g.

41. Old Tory

42. Mandela's land: Abbr.

43. Work without ___

44. Rose-red dye

46. Where the worm turns

50. Overwhelmingly

52. Verizon FiOS, e.g., for short

55. Org. with an annual televised awards ceremony

56. Fill in the blank with this word: "BBC : Britain :: ___ : Italy"

57. Ryder Cup scoring method

60. Either way, Cupid recognized my pain

63. Weather info: Abbr.

64. Showed to the foyer

65. Tennis whiz

66. Snake sound

67. Talks little

68. Times Square sign

DOWN

1. Shade of brown

2. The English translation for the french word: oindre

3. With or without an "H" at the end, this is a title for a Muslim teacher; a high-ranking one may be an ayatollah

4. Cosmos star

5. Fill in the blank with this word: "___ the custom (traditionally)"

6. With a diameter averaging about 7 miles, it's the smaller of Mars' 2 moons

7. Small-circulation publication for fans

8. Ticker tests, for short

9. Michigan's ___ Canals

10. Split up

11. Prosodic pauses

12. Gosh, British-style

13. Stock units: Abbr.

18. Three ___ match

19. Super Bowl XV champs

24. Persist

25. Memo abbr.

26. Washington and ___ University

28. Hurricane-tracking agcy.

30. Civilian clothes

31. Overseas bar deg.

34. Creator of Aslan and the White Witch

35. See 24-Across

37. Vitamin bottle info

38. Fill in the blank with this word: "___ mots"

39. Tears?

40. WWII intelligence org

41. Word often heard in triplicate

45. Lamebrain

47. Fill in the blank with this word: ""One ___ Jump""

48. Turkish inn

49. Tiramisu features

51. Tobacco kilns

53. Some hospital procedures

54. Chem. class measures

57. Sound of an air kiss

58. Sony subsidiary

59. Warm-up for the college-bound

60. Workup locales: Abbr.

61. Retailer with stylized mountaintops in its logo

62. Camera operator's org.

PUZZLE 25

1	2	3	4		5	6	7	8		9	10	11	12	13
14					15					16				
17				18					19					
20								21						
22							23							
			24		25	26			27		28	29	30	31
32	33	34		35				36			37			
38			39						40	41				
42					43						44			
45				46	47		48			49				
			50		51				52		53	54	55	
56	57	58	59					60	61					
62						63								
64					65					66				
67					68					69				

ACROSS

1. Unscramble this word: dere

5. Steps down to a river, in India

9. William who wrote "Half Mile Down"

14. Trix alternative?

15. Green-skinned dancing girl in a "Star Wars" film

16. This Houston energy company was delisted from the NYSE in 2002; later many of its execs pleaded the Fifth

17. 1997 in-flight movie?

20. Trig angle

21. The Wright brothers, e.g.

22. To look, in Leipzig

23. Voiced by John Cleese, King Harold of "Shrek" fame was originally one of these small creatures

24. "Battlestar Galactica" commander

27. Fill in the blank with this word: ""Lo and ___!""

32. Unrealized 60's Boeing project

35. Miyoshi ___, Best Supporting Actress winner for "Sayonara"

37. Dynasty in which Confucianism and Taoism emerged

38. Egotist's favorite person?

42. Top Tatar

43. Mixer maker

44. Range part: Abbr.

45. Lug: Var.

48. Fill in the blank with this word: ""___ sorry!""

50. Mil. unit below a division

52. Toothpaste with a spokesbeaver

56. It may help you keep your place

60. Students with personal guides

62. Cleopatra's last words

64. Fill in the blank with this word: ""Bonne ___!" (French cry on January 1)"

65. 'Houston, ___ had a problem'

66. Words often before a colon

67. Wrapper stat.

68. ___ Eyes' (1975 #2 hit)

69. Political cartoonist Thomas

DOWN

1. Manet contemporary

2. "More!"

3. Violinist/bandleader ___ Light

4. Waters parted in Exodus

5. Fill in the blank with this word: ""How's it ___?""

6. Swimming ___

7. Fill in the blank with this word: "Dryden's "___ for Love""

8. Small drum

9. The English translation for the french word: s˚r

10. Within: Prefix

11. Italian emporium ending

12. Until 1990 it was the capital of West Germany

13. Some specialize in elec.

18. operated by a mechanism

19. Worrisome remark by a surgeon

23. Puts on an act

25. What a guitar may be hooked up to

26. Prefix with drama

28. Big jet

29. Fill in the blank with this word: ""It's us against ___""

30. Fill in the blank with this word: "Dennis ___ and the Classics IV (1960s-'70s group)"

31. one of five children born at the same time from the same pregnancy

32. Library section

33. "Zip it"

34. Wild duck

36. Whit

39. Work without ___

40. Opposite of masc.

41. Noble family of medieval Italy

46. Ready already

47. Farmyard female

49. Doesn't decline

51. Stomach sound

53. UnitedHealth rival

54. Moves toward

55. Up to the present

56. Fill in the blank with this word: "Allan ___, "Sands of Iwo Jima" director"

57. Fill in the blank with this word: ""Die Frau ___ Schatten" (Strauss opera)"

58. Fill in the blank with this word: "Eugene ___, hero of "Look Homeward, Angel""

59. Worry

60. X years before Hastings

61. The Bible's Garden of ___

63. One-eighty

PUZZLE 26

1	2	3	4		5	6	7	8		9	10	11	12	13
14					15					16				
17					18					19				
20				21				22						
		23				24					25	26	27
28	29	30			31		32			33				
34				35		36				37				
38			39				40	41	42					
43						44					45			
46			47	48	49		50			51				
52			53			54		55						
		56	57				58				59	60	61
62	63				64				65					
66					67				68					
69					70				71					

ACROSS

1. Year that Clement XI became pope

5. Fill in the blank with this word: “”Veni, vidi, ___””

9. International Olympics chief Jacques

14. Seal's opening?

15. Some Red Cross work: Abbr.

16. Outsider, in Hawaii

17. Fill in the blank with this word: “”The very ___!””

18. Midnight alarm giver

19. Prussian lancer

20. Rich pie filling

23. Year in the life of Leif Ericson

24. This word for a dolt was the first name of Rube Goldberg's comic strip hero McNutt

25. Suffix with morph-

28. Unscramble this word: apsnios

32. Gripes

34. Thomas Moore poem “___ in the Stilly Night”

35. Fill in the blank with this word: “”Last one ___ a rotten egg!””

37. Meth., e.g.

38. -

43. "Goldberg Variations" composer

44. Actress Rowlands

45. End of many an E-mail address

46. Perfectly restored

50. Lions or Bears

52. Zine staff

53. Undecided, you might say

55. Fill in the blank with this word: "BBC : Britain :: ___ : Italy"

56. Person who looks exactly like another

62. Fill in the blank with this word: "___ liberales (studies at universidades)"

64. Loc. of some devils

65. Their, in Munich

66. Fill in the blank with this word: ""___ c'est moi""

67. Trix alternative?

68. Simplest of choices

69. Legless creatures

70. Rick with the 1976 #1 hit "Disco Duck"

71. Ovid's family name

DOWN

1. Wing, say

2. What ___ tell you?'

3. Street: Abbr.

4. Gorges

5. Competing narratives

6. Folk singer Burl

7. West Indies native

8. Colorful summer treat

9. Sensuous ballroom dance

10. Where Pearl City is

11. Good neighbor policy

12. The English translation for the french word: acide ?-linolÈnique

13. They dug his grave ___ where he lay': Sir Walter Scott

21. Five, on a gunslinger's gun

22. Neighbor of Ger.

26. Peak of NE Greece

27. Value

28. Toeless creature in an Edward Lear verse

29. Yellow

30. Treats unfairly

31. Wellness org.

33. You might take stock in it: Abbr.

36. Wilt

39. Fill in the blank with this word: ""Thanks for ___ Memory""

40. Fill in the blank with this word: "Basketball's ___ Elmore"

41. Puts a border around

42. Judaism : kosher :: Islam : ___

47. Outlawed blasts

48. Strauss's "___ Heldenleben"

49. Made damp

51. Painter surnamed Vecellio

54. Steak ___

57. Fill in the blank with this word: "___ as a doornail"

58. Fruit holder

59. Fill in the blank with this word: ""___," said Tom haltingly"

60. Some flawed mdse.

61. Biblical peak

62. Pie ___ mode

63. Union ___

PUZZLE 27

```
1   2   3   4   █   5   6   7   8   █   9   10  11  12  13
14          █   15          █   16
17          █   18          █   19
20          21          22      █   23
█   24          █   25      █   26  █
27  28  29      █   30  █   31          32  33  34
35      █   36      █   37      █   38
39      40          █   41  42
43      █   █   44          █   █   45
46          47  48  █   49      50
█   51          █   52  █   53      █
54  55  56  █   57          58      59  60  61
62      63  █   64          █   65
66          █   67          █   68
69          █   70          █   71
```

ACROSS

1. Wait ___!' ('Hold on!')

5. New York's Washington Square ___

9. Midsection, informally

14. Ticket choice

15. Scientology's ___ Hubbard

16. German indefinite article

17. Zip

18. Fill in the blank with this word: "Emporio ___"

19. 10,000,000 rupees, in India

20. Sloven in the coven?

23. Fill in the blank with this word: "___-a-Mania, candy collectors' convention since 1991"

24. X-ray spec?

25. Skittles

27. Hot-dog

31. systematic (similar term)

35. Woe ___' (popular grammar book)

36. Course for course preparers

38. Fill in the blank with this word: "___'clock"

39. Merely support

43. St. ___ (malt liquor brand named after an Irish nun)

44. Heart of France

45. Suffix with ball or bass

46. Immature swimmer

49. Colts' former home

51. Salt-___ (rap trio)

53. Fill in the blank with this word: "Bill and ___"

54. Fill in the blank with this word: "___ Hatter"

57. Angle

62. Fill in the blank with this word: "___ One (indoor kart racing)"

64. Punnily titled 1952 quiz show "Up to ___"

65. What I will follow

66. Illumination of manuscripts, and others

67. Yeats's land

68. Words with a nod

69. Fill in the blank with this word: ""What have ___?""

70. Slave to detail

71. Hall-of-Fame college swimming coach ___ Thornton

DOWN

1. Hebrew letter

2. Songs for one

3. Alike: Fr.

4. Prefix with -fugal

5. Know nothing, so to speak

6. Got ___ deal (was rooked)

7. Samurai without a master

8. Shopper for woolen goods?

9. Some of Moby's music

10. Onetime TWA rival

11. The L train?

12. Fill in the blank with this word: "A ___ technicality"

13. Santa ___ (neighbor of Lompoc, California)

21. Kind of yoga

22. The Reds, on scoreboards

26. Young lady of Sp.

27. Words to the maestro

28. Fill in the blank with this word: "Carne ___ (roasted meat dish)"

29. Amplified

30. They go around on reservations

32. Fill in the blank with this word: "Artoo-___"

33. Sci. of insects

34. TV's "20/20" creator

Arledge

37. Umberto ___, author of 'The Name of the Rose'

40. Locker room shower?

41. Giver of relief

42. Fill in the blank with this word: "___ Good Feelings"

47. Unscramble this word: ppooes

48. Uncle ___ of "Seinfeld"

50. Attacked with zeal

52. Fill in the blank with this word: ""For want of ___ …""

54. Year the oldest college in the Americas was founded, in Mexico City

55. Youngest player to join the 500-homer club

56. Fill in the blank with this word: ""I can only ___ much""

58. Weight not charged for

59. Then preceder

60. Elbe tributary

61. Stimulate, as the appetite

63. Range part: Abbr.

PUZZLE 28

ACROSS

1. Goes by foot, with "it"

6. Weight-bearing bone

11. Three ___ and One DJ' (Beastie Boys song)

14. Suffer ___ worse than death

15. One of a Western political family

16. Suffix with torrent

17. Result of sitting on a court bench too long?

20. Wreck-checking org.

21. Court opening

22. of or relating to or characteristic of Genoa or its inhabitants

26. Renowned chair designer

27. Year in Pelagius I's papacy

28. TV's ___ Network

31. The item seen here, or the taste it might have if made with lemons

32. Honest ___

33. Fault line?

37. Story written by 38-Across

42. Principal player in "Grease"

43. Fill in the blank with this word: "Ad ___ (relevant)"

44. Staff differently

47. WSW's reverse

48. Three-part ordeal for H.S. students

49. Oafish

51. Warns maliciously

55. Apollo as god of the sun

58. Fill in the blank with this word: "___-Day"

59. Robert Rubin's successor as Treasury Secretary

64. One in the charge of un instituteur

65. This garlic-flavored mayonnaise from Provence is popularly served with fish

66. Fill in the blank with this word: "___ acid"

67. Tenth letter of the Hebrew alphabet

68. Betelgeuse, for one

69. Woodworking fasteners

DOWN

1. Book before Zephaniah: Abbr.

2. U.S. ___

3. Western treaty grp.

4. U.S. bullion site

5. Fill in the blank with this word: ""Must-___" (NBC slogan)"

6. Marcus or Winfred of the N.F.L.

7. Ore-___ (frozen food brand)

8. Debutantes' affairs

9. Fill in the blank with this word: ""Winnie ___ Pu""

10. Emmy winner Woodard

11. Nadirs

12. Summer sign

13. Least ingenuous

18. Taoism founder Lao-___

19. The English translation for the french word: avatar

22. Fill in the blank with this word: "___ : hello :: hooroo : goodbye"

23. Wittenberg's river

24. Forty-___

25. Swindles

29. Took to court

30. Konrad Adenauer, Der ___

33. XI years before the Battle of Hastings

34. Unusual shoe spec

35. Mussorgsky's 'Pictures ___ Exhibition'

36. Wit's end?

38. Many an agent

39. They often precede la's

40. Fill in the blank with this word: ""I ___ Song Go Out of My Heart""

41. Vent, in a way

44. Robert ___

45. Surround with radiance

46. Fill in the blank with this word: "Dog : barked :: cat : ___"

48. Hit hard, as brakes

50. Letter-shaped bridge support

52. Response to a general question?

53. Australian singer Christine

54. Fill in the blank with this word: ""Please ___" (invoice stamp)"

56. Les …tats-___

57. Fill in the blank with this word: "___-free"

60. She, in S

61. Web site address ending

62. Slowing, in mus.

63. Barges

PUZZLE 29

ACROSS

1. Robt. E. Lee, e.g.

5. With 69-Across, 1930s-'50s bandleader

9. U.K. equivalent to an Oscar

14. Fill in the blank with this word: "___ fide"

15. Ventura County's ___ Valley

16. Fill in the blank with this word: "D'___ (according to: Fr.)"

17. Up and ___!'

18. Norse goddess of fate

19. Smock

20. What kvetches do at grocery stores?

23. Training ___

24. Vex

27. Visitors to the Enterprise

30. See eye to eye about

34. N.L. team, on scoreboards

35. Prophet who predicted the destruction of Nineveh

37. TV co-star of Richard Belzer

38. The English translation for the french word: miso

39. Query in a punny 39-Across poem about eating out

42. Finish this popular saying: "If wishes were horses, beggars would_____."

43. Where William the Conqueror is buried

44. TV's "20/20" creator Arledge

45. Confounded

46. Site of the Bay of Whales

48. James Fenimore Cooper's "___ Myers"

49. Swiss developmental psychologist

51. Super Bowl winner more than 50% of the time: Abbr.

53. Nerd's essence?

60. Take ___ (look)

62. The ___-Neisse Line

63. Fill in the blank with this word: ""___ Simple Man" (#1 Ricky Van Shelton song)"

64. Treeless tract

65. Fill in the blank with this word: "___ oak"

66. Madrid month

67. Miniseries segment

68. Side-channel, in Canada

69. The English translation for the french word:

thÈorie des ensembles

DOWN

1. Onetime big inits. in car financing

2. I could ___ horse!'

3. What an A is not

4. Some soot

5. Writer Susan

6. Fill in the blank with this word: "Do ___ on"

7. Fill in the blank with this word: ""Matilda" star ___ Wilson"

8. Like some vases

9. Whalebone

10. Unscramble this word: atarp

11. Play and film about a noted 1977 series of interviews

12. The Missing Drink : High ____ rose

13. Winter hrs. in Bermuda

21. Mild-flavored seaweed in Japanese cuisine

22. Thou

25. Essential amino acid

26. Mugged

27. Swaddle

28. Fill in the blank with this word: "Bernstein's "Trouble in ___""

29. Is somewhat remorseful

31. The Beatles' meter maid and others

32. Repeat calls?

33. Women's dress sizes

36. Yorkshire river

38. Tropical fruit

40. Tartan hose wearers

41. "Mr. Belvedere" co-star

46. Smell like

47. Massive, in Marne

50. The treaty of this city ending the War of 1812 returned all lands captured by Britain & the U.S. during the war

52. Rail supports

54. Fill in the blank with this word: "___ and aahs"

55. Tempura ___ (Japanese dish)

56. Sir Peter ___, painter of British royalty

57. Wind down

58. TV's 'How ___ Your Mother'

59. Town NNE of Santa Fe

60. Yodeler's perch

61. Mao's mil. force

PUZZLE 30

ACROSS

1. D.O.E. part: Abbr.

5. Pull out of ___ (produce suddenly)

9. Send-offs of a sort

14. Fill in the blank with this word: "___ Verde National Park"

15. Fill in the blank with this word: "___ of Venice"

16. Fill in the blank with this word: "___ Perot"

17. Torment

18. Bond villain in "Moonraker"

19. Softens

20. *Does a hostler's work

23. Kind of a drag

24. First Chinese dynasty

25. Secure online protocol

28. Withhold praise for certain football linemen?

31. Some sporty cars

34. Waiting area announcements, briefly

35. Fill in the blank with this word: ""I earn that ___": As You Like It""

36. They're straight

38. 'Worcester, get thee gone' speaker

41. "... abridging the freedom of speech, ___ the press ..." (Bill of Rights)

42. To do this is to stare at someone desirously

43. The Unsers of Indy

44. Locate cookware components?

49. The English translation for the french word: c.‡.c.

50. The English translation for the french word: Tanguy

51. Folk/country singer Griffith

54. President Ford stared fiercely

57. Hong Kong neighbor: Var.

60. Look to ___ troublous world': 'Richard III'

61. Smear with wax, old-style

62. "Dang!"

63. Quod ___ faciendum

64. Wail

65. Fill in the blank with this word: ""Do as I say, not ___""

66. Go on a vacation tour

67. Use a knife

DOWN

1. CPR pros

2. Wrapper stat.

3. Fill in the blank with this word: "___ Lauder"

4. Like some bonds

5. Totals

6. Fill in the blank with this word: "___ d'oeuvre"

7. Mystery writers' awards

8. Austin school, informally

9. "Riverdance" composer Bill

10. Principal river of Armenia

11. WBA wins

12. Yacht's dir.

13. Worrying sound to a balloonist

21. Tootler

22. Topper

25. Words said upon departure

26. These gastropods are sometimes fed aromatic herbs to give them a special savor

27. Modern viewing options, for short

29. The Mormons, initially

30. Listen: Sp.

31. Unscramble this word: ohtsg

32. University of Maryland, informally

33. The English translation for the french word: piquÈ

37. Red ___ (young amphibian)

38. Elev.

39. Stretch out

40. Want to ride an elephant in this country? Head for the Gokarna Safari Park east of Katmandu

42. Workers with horse sense?

45. Breath: Prefix

46. Whitman's "A Backward Glance ___ Travel'd Roads"

47. Ribs

48. They're often taken to kindergarten

52. Angler's basket

53. Middle of a famous palindrome

54. Showy trinket

55. "My stars!"

56. The English translation for the french word: tache

57. Wharton grad

58. Cries of regret

59. The English translation for the french word: CGI

PUZZLE 31

1	2	3	4	5		6	7	8	9		10	11	12	13
14						15					16			
17						18					19			
20					21						22			
		23					24		25					
26	27	28		29		30	31		32					
33			34		35			36		37		38	39	40
41			42						43					
44					45						46			
		47		48		49			50		51			
52	53	54				55			56		57			
58				59		60	61				62	63	64
65				66					67					
68				69					70					
71				72					73					

ACROSS

1. Sickly institution of the 80's

6. Sound of a leak

10. Where chamois and snow leopards live: Abbr.

14. Get ___ reception

15. The U.N.'s Kofi ___ Annan

16. Whose woods these ___ think...': Frost

17. Fill in the blank with this word: ""___ Doone""

18. Not masc. or fem.

19. Though spelled differently than her BFF, Ms. King, this is Oprah's middle name

20. Drug source [right only]

22. Seraph of S

23. Mai ___

24. Fill in the blank with this word: "Best-___ (dog show category)"

26. The English translation for the french word: moi

29. The English translation for the french word: plaie

32. Variety of whale

33. Gold-medal Olympic swimmer ___

Torres

35. ___ Kalugin, former K.G.B. general with the 1994 book "Spymaster"

37. Schnozz

41. They keep airliners aloft

44. Fill in the blank with this word: "___ a fox"

45. Xanadu river

46. Lacoste of tennis

47. U.K. heads

49. Ja and da

51. Kind of I.R.A.

52. Hot

56. Stock units: Abbr.

58. Go on a vacation tour

59. Foreboding cartoons?

65. Wasn't straight

66. Fill in the blank with this word: "___ wheelie"

67. XM ___

68. Work in the cutting room

69. Keto-___ tautomerism (organic chemistry topic)

70. Wooded

71. Fasten on

72. Cub #21 of the 1990s-2000s

73. Straws in the wind

DOWN

1. Eastern European pork fat dish

2. Reynolds film "Rent-___"

3. It's the 4-letter term for the thin sheets of dried seaweed in which sushi is wrapped

4. Sinker

5. Foals : horses :: crias : ___

6. Fill in the blank with this word: ""Mens sana in corpore ___""

7. One ___ at a time

8. Short and thick

9. Part of Dionysus' entourage

10. What the starts of the answers to the eight starred clues are

11. Real downer, for short?

12. Whinny

13. Proctor-___ (small appliance maker)

21. Molly of Yiddish theater

25. Kipling's "___ we forget!"

26. Old English letters

27. The English translation for the french word: prison

28. The English translation for the french word: Orly

30. The English translation

for the french word: algue

31. Unscramble this word: reply

34. Got comfortable with

36. Yawn

38. ___ Island (location near Portland, Maine)

39. One less than une douzaine

40. Encouraging sign

42. This ___ quest...' ('The Impossible Dream' lyric)

43. What foxhounds try to catch

48. Hit the ___

50. Plot element?

52. Fill in the blank with this word: "___ of Langerhans"

53. [See title, and proceed]

54. Try to ___ my way...'

55. See 81-Across

57. Unctuous flattery

60. Works of Homer

61. "The Lion King" lion

62. Fill in the blank with this word: "___ fixe"

63. Hundred, in Ju

64. Lays down the lawn

PUZZLE 32

ACROSS

1. Tennis whiz

5. Tropical tuber

9. Play by a different ___ rules

14. Prix de ___ de Triomphe (annual Paris horse race)

15. Fill in the blank with this word: ""___ le roi!""

16. Where drachmas were once spent

17. Quote, part 2

20. The English translation for the french word: anorak

21. Fill in the blank with this word: "___ Park, home for the Pittsburgh Pirates"

22. Words with a nod

23. Whistler, at times

24. In 1986 U.S. Steel changed its name to this 3-letter one

26. Urge

28. Fill in the blank with this word: "___-Pei (dog)"

30. Fill in the blank with

this word: ""Pretty ___""

34. Usher's offering

37. Fill in the blank with this word: "___-Day"

39. The "L" of A.F.L.-C.I.O.

40. Courage seeker in a 1939 film

44. Watchers

45. Tic-tac-toe choice

46. Volkswagen model

47. Value

49. What ___ thinking?'

51. Like llamas

53. The English translation for the french word: OMC

54. Fill in the blank with this word: "Doo-woppers ___ Na Na"

57. Suffragist Carrie Chapman ___

60. Fill in the blank with this word: ""Don't tell ___ can't ...!""

62. Yanks' org.

64. One who has practiced his hitting skills

67. Town council president, in Canada

68. Water barrier

69. Speed skater Apolo Anton ___

70. More desertlike

71. Katharine's role in "Adam's Rib"

72. Wilts

DOWN

1. ___ boy

2. Fill in the blank with this word: "Chili con ___"

3. Fill in the blank with this word: "___ Good Feelings"

4. Stereo syst. component

5. Place to plug a new book, maybe

6. Washboard ___

7. Woodworking tool

8. "Sesame Street" tune, with "The"

9. On the Road' narrator ___ Paradise

10. The best you can be, Freudian-style

11. Tugboat services

12. The English translation for the french word: orle

13. Tammy ___ of 1970s-'80s TV

18. Waters, informally

19. Old English bard

25. Popular antianxiety drug

27. Unite formally

29. Arrange into new lines

31. It's not ___ deal'

32. The English translation for the french word: suie

33. About

34. To ___ (unerringly)

35. John ___-Davies of the "Lord of the Rings" trilogy

36. Track ___

38. Play to ___ (tie)

41. Outside-the-box

42. Douay prophet

43. Fill in the blank with this word: "___.com (e-mail address for the Coast Guard)"

48. They say this lady will "coax the blues right out of the horn" & "charm the husk right off of the corn"

50. Seat of Allen County, Kansas

52. Canio's wife in "Pagliacci"

54. Fill in the blank with this word: ""I Am ... ___ Fierce," #1 Beyonc"

55. When doubled, a former National Zoo panda

56. B and O figures: Abbr.

57. They play in front of QBs

58. Answer to the riddle "Dressed in summer, naked in winter"

59. Six-foot vis-

61. Fill in the blank with this word: ""___ to please""

63. Those, to Tom

65. Fill in the blank with this word: "___ Tamid (synagogue lamp)"

66. Dog doc

PUZZLE 33

ACROSS

1. Yacht spot

5. Fill in the blank with this word: "___ cheese"

9. What you may call it?

14. QB Tony

15. Lacoste of tennis

16. Fill in the blank with this word: "Caput ___ syndrome (arm problem)"

17. Runs a tab

18. Unbending

19. Video game pioneer

20. Algae color

23. Rock's ___ Jovi

24. Fill in the blank with this word: ""___ only""

25. Just one little bite

27. Soup brand

30. Take turns?

32. Yamaha product, briefly

33. Best at putting things away

36. Not ___ many words

39. Pigeon-___

41. "Of course!"

42. Spittoon sound

43. Bank-to-bank transactions: Abbr.

44. Actresses Eleniak and Alexander

46. Taken ___

47. Spade portrayer

49. Range features

51. Tilts

53. Their days are numbered

55. Fill in the blank with this word: "0% ___"

56. *Crunchy item at a salad bar

62. Ibsen's Gabler

64. Sleep like ___

65. First Chinese dynasty

66. Lend ___ (pay attention)

67. Loc. of some devils

68. Ogler

69. Thin iPods

70. Suburb of Paris

71. Work over

DOWN

1. Where a queen may be crowned

2. To stand in the center of this state, go 5 miles northeast of Ames & stand there--how exciting...

3. Retired, as a prof.

4. More upbeat

5. Fill in the blank with this word: "___ Lauder"

6. Burdens

7. France's Cote d'___

8. Unspoken

9. Portions

10. Last: Abbr.

11. One way to vote

12. This card deck's Minor Arcana has 14 cards in each suit; a page is between the 10 & jack

13. The English translation for the french word: seine

21. Singer-songwriter Laura

22. Singer Kitt

26. On one's toes

27. William of TV's "The Greatest American Hero"

28. Like many office jobs

29. the surface soil that must be moved away to get at coal seams and mineral deposits

30. See

31. Talk, talk, talk: Var.

34. Trepidation

35. Fill in the blank with this word: "Basher ___, one of the eleven in "Ocean's Eleven""

37. Resist cheerfulness

38. The Van Eyck bros. weren't the 1st to use these paints, though sometimes credited with it; they date back much earlier

40. U.K. military medals

45. Hydros : England :: ___ : U.S.

48. Sporty wheels

50. Williams in the water

51. Shah ___ (Taj Mahal builder)

52. Fill in the blank with this word: "___ can of worms"

53. Unscramble this word: ocrss

54. The English translation for the french word: pygmÈe

57. They follow so

58. They, in S

59. Fill in the blank with this word: ""For ___ sow...""

60. Wasn't straight

61. Syrup brand

63. Fill in the blank with this word: "1957 Physics Nobelist Tsung-___ Lee"

PUZZLE 34

1	2	3	4	5	■	6	7	8	9	10	■	11	12	13
14					■	15					■	16		
17				18							■	19		
20			■	21			■	22		23				
24			25			26	■	27				■		
■		28			■	29	30				■	31	32	
33	34	35			■	36			■	37				
38			■	39				■		40				
41		42	■	43				■	44	45				
46			47			■	48				■			
■	49				■	50	51			■	52	53	54	
55	56			■		57			■	58				
59			■	60	61	62			63					
64			■	65			■	66						
67			■	68			■	69						

ACROSS

1. Lulu

6. Plover named for its call

11. Pitcher Robb ___

14. Violists' places: Abbr.

15. Yoke

16. Wall Street org.

17. Farmer's overalls?

19. Opposite of bellum

20. Fill in the blank with this word: ""For ___ a jolly Ö""

21. Fill in the blank with this word: "Debussy's "Air de ___ ""

22. Went to Wendy's, say

24. Old dancing duo

27. Fill in the blank with this word: "___-Ration (dog food)"

28. ___ House and Museum (Baltimore attraction)

29. The English translation for the french word: Èmergent

33. Strawberry, once: Abbr.

36. The English translation for the french word: estrade

37. Zoom

38. The English translation for the french word: ennemi

39. Unit used in electromagnetism

40. Wichita-to-Omaha dir.

41. Width's opposite: Abbr.

43. The English translation for the french word: razzia

44. Singer-actress Janis

46. Exigency

48. Yahoo! competitor

49. Cyclops' feature

50. Overshadowed

55. Speak sharply to

57. Years on end

58. Fill in the blank with this word: "___ Friday's"

59. Whitman's "A Backward Glance ___ Travel'd Roads"

60. ... in 13-Down

64. Writer Rita ___ Brown

65. Unscramble this word: atarp

66. Grande ___ (Qu

67. She, in S

68. Relatively cool red giant

69. Parts of an udder

DOWN

1. 1993 Danny Glover film based on a Percy Mtwa play

2. Mystery writer Gardner et al.

3. Fill in the blank with this word: "At ___ (with repercussions)"

4. Beat and how!

5. 'The Waste Land' poet

6. Cocoon dwellers

7. Spanish queen until 1931

8. Victory

9. Quip, part 2

10. The English translation for the french word: testeur

11. Subject of an Emil Ludwig biography

12. Missing name in the tongue twister 'I saw ___ sawing wood ...'

13. Word to waiters

18. To laugh, to Lafayette

23. Some specialize in elec.

25. Ursine : bear :: pithecan : ___

26. Vacation destination

30. Temperate

31. Fill in the blank with this word: "Da ___, Vietnam"

32. Fill in the blank with this word: "Family ___"

33. Can. province

34. Wise one

35. Radio station's coverage, perhaps

36. "My stars!"

39. Mountain view

42. The hip of the dog-rose

44. Indicate

45. Yodeler's perch

47. Striped antelopes

48. Golden ___ (century plant)

51. Middle: Prefix

52. Lady Liberty garb

53. Snowy ___

54. Dutch sights

55. Fill in the blank with this word: """___ nerve!""

56. Broten of hockey fame

61. Was on the bottom?

62. Fill in the blank with this word: "___-la-la"

63. Corduroy ridge

PUZZLE 35

ACROSS

1. Mrs. Lincoln's maiden name

5. Year in John XVIII's papacy

9. Singer McEntire and others

14. Singer nicknamed 'The Jezebel of Jazz'

15. Gear teeth

16. Filled (with)

17. Way it's done

18. -

19. Stuffed ___ (kishke)

20. Bullies

23. Scott Joplin's "Maple Leaf ___"

24. In general

25. Revenuers

26. Young haddocks

29. Shakespeare's ___ of Salisbury

31. Newspaper's ___ page

32. Cleopatra's last request?

37. Where to see a

mummy: Abbr.

38. Comment after a difficult decision

40. Fill in the blank with this word: ""Days of ___ Lives""

41. It's hard to understand

43. Woes of the world

44. Word before face or heart

45. Small bay

47. Vingt-___ (blackjack)

49. Patella

53. Fill in the blank with this word: "___ sgt. (police rank)"

54. Survivors' concerns

58. The English translation for the french word: sec

60. Writer LeShan and others

61. Ones in charge: Abbr.

62. Twisted humor

63. '60s protest / Skip, as a dance

64. Fill in the blank with this word: ""___ kleine Nachtmusik""

65. German indefinite article

66. Fill in the blank with this word: "Eye ___"

67. ___ le roi!' (French Revolution cry)

DOWN

1. Fill in the blank with this word: "___ of the Unknowns"

2. Whiff

3. Woodworking groove

4. Natural tint source

5. Radio couple at 79 Wistful Vista

6. relating to the palm of the hand or the sole of the foot

7. Marvin Gaye's "Can ___ Witness?"

8. Fill in the blank with this word: "Dan ___, former N.B.A. star and coach"

9. Some mail designations: Abbr.

10. We'll teach you to drink deep ___ you depart': Hamlet

11. Air show maneuver

12. Sportscaster Rashad

13. Some wild parties

21. Underwear initials

22. Like non-oyster months

25. To and ___

26. Fill in the blank with this word: ""___ nerve!""

27. The "brains" of 58-Down

28. Unscramble this word: slntoeuroi

30. Where the outboard motor goes

32. Fill in the blank with this word: ""___ was saying Ö""

33. Year that Dionysius of Halicarnassus is believed to have died

34. Fill in the blank with this word: "___ green"

35. Uncooperative one

36. Fill in the blank with this word: ""The even mead, that ___ brought sweetly forth ...": "Henry V""

38. What "y" might become

39. Wilderness walks

42. Whole lot

43. Spears

45. Scores 100 on a test

46. Kangaroo ___

47. 1983: "___ and the Cruisers"

48. Fill in the blank with this word: "___ Clark who sang "Poor, Poor Pitiful Me""

50. Weapons check, in brief

51. Fill in the blank with this word: ""Embraced by the Light" author Betty J. ___"

52. The Louvre's Salles des ___

54. Fill in the blank with this word: "___ Grand (supermarket brand)"

55. Marked, as a questionnaire box

56. About

57. Fixed at an acute angle

59. WSW's reverse

PUZZLE 36

1	2	3	4	5		6	7	8	9		10	11	12	13
14						15					16			
17					18						19			
20								21		22				
			23			24				25				
26	27	28	29					30	31					
32						33						34	35	36
37					38	39					40			
41				42						43				
		44					45	46						
47	48	49				50								
51				52	53		54				55	56	57	58
59				60		61								
62				63					64					
65				66					67					

ACROSS

1. Fill in the blank with this word: """___ to recall "

6. Twice tetra-

10. Fill in the blank with this word: "___ McAn shoes"

14. You've got the wrong guy!'

15. Essential company figure

16. Philosopher David

17. Like aprons, at times

19. They're often bitter

20. Words from a would-be protégé

21. Turn over again

23. Lummoxes

25. The second part missing in the author's name ___ Vargas ___

26. Famous place with a hint to this puzzle's theme

32. Skeleton components

33. Wine taster's adjective

34. Meditation sounds

37. Heroin, slangily

38. "Hamlet" courtier

40. "That was close!"

41. Elev.

42. Ural River city

43. Fill in the blank with this word: "2003 Nick Lachey hit "___ Swear""

44. 2001 50-Across nominee

47. Mirabile ___ (wonderful to say)

50. Shakespearean king

51. Library section

54. Fire sources

59. I Lost It at the Movies' writer Pauline

60. Some clerks

62. One way to race

63. Sound system brand

64. Tennessee ___ Ford

65. What knows the drill, for short?

66. Dilbert co-worker

67. Marie Antoinette, e.g.

DOWN

1. Research facility: Abbr.

2. 1982 title role for Meryl Streep

3. Ici ___ (here and there, to Th

4. Old Apple product marketed to schools

5. Ways

6. Political activist James known for undercover videos

7. Nav. leader

8. Six-foot vis-

9. French wave

10. Fill in the blank with this word: "Buddy Holly's "___ Be the Day""

11. 1950s fad item

12. Biblical dry measures

13. Legendary Washington hostess Perle ___

18. Yao Ming teammate, to fans

22. Wow

24. View coral reefs, maybe

26. Talk effusively

27. Cockney greeting

28. Sci. course

29. This 3-letter word means to pester or badger; it can also refer to a worthless horse

30. Mai ___

31. QB Manning

34. Fill in the blank with this word: "___ Valley Conference"

35. The Bible Tells ___'

36. Triathlon leg

38. A Thing ___' (Beach Boys song)

39. Secure online protocol

40. Sentence part: Abbr.

42. Fill in the blank with this word: "___ probandi"

43. First-rate

44. Slanted type

45. Trim

46. Rabbit ___

47. What is the capital of this country - Senegal

48. Give ___!' ('Try!')

49. Traction provider

52. Fill in the blank with this word: "___ noche (tonight, in Tijuana)"

53. Writer's supply: Abbr.

55. Clinton's #2

56. Present opener?

57. Teutonic turndown

58. Snick-or-___

61. Philosopher ___-tzu

PUZZLE 37

ACROSS

1. Senator Jake in space

5. Carroll:auth.::Tenniel:___

10. Two-out actions: Abbr.

13. Queen in a long-running comic strip

15. Capital city about an hour by plane from Miami

16. Rock's ___ Speedwagon

17. Delectable

19. Year that Augustus exiled Ovid

20. Rake over the coals, say

21. Tower supports

23. Japanese prime minister Taro ___

24. This ancient Hebrew measure equal to about 2 quarts sounds like a synonym for "taxi"

25. Time for a coffee break, maybe

27. What Santa's elves rode when the sleigh broke down?

31. Hat designed Lilly

34. Fill in the blank with this word: ""___ a Putty Tat" (Friz Freleng short)"

35. Fill in the blank with this word: "___-Locka, Fla."

36. You might stick a knife in it

37. Witch of ___

39. Rheinland residence

40. U.S.N. officers

41. Fill in the blank with this word: "Amo, amas, ___"

42. U.R.L. opener indicating an additional layer of encryption

43. Bygone children's book character

47. Roman sandal

48. Old bird

49. Fill in the blank with this word: "China's Chairman ___"

52. Crocus and gladiolus

54. Halite

56. U.S. Open hit

57. Something a company won't reveal

60. Fill in the blank with this word: "___ system"

61. Fill in the blank with this word: "Be ___ (constantly complain)"

62. "Hawaii Five-O" name

63. Wooed very well

64. Nancy's opposite number, once

65. Per ___ (daily)

DOWN

1. Nero's successor

2. The English translation for the french word: alias

3. Paste-up piece

4. Wreck-checking org.

5. Spirits that victimize the sleeping

6. While you're enjoying the mountain scenery you might spot a tarn, a small one of these left by a glacier

7. Long-running B'way musical seen by couples?

8. Swabs' grp.

9. Great Lakes port of entry

10. Hunk

11. The Premier Rose Diamond is one of the world's largest diamonds cut in this fruit shape; it was sold for $10 million

12. Lays down the lawn

14. The Story of Alexander Graham Bell' star

18. Rent-___

22. Fill in the blank with this word: "___-Rhin (Strasbourg's department)"

25. Torment

26. Webzine

27. John of the Harold & Kumar films

28. Some bank deposits

29. Fill in the blank with this word: ""___ and away!""

30. Words of disrespect

31. Tot's tote

32. Some voices

33. Song title followed by the lyric "Lovers say that in France"

37. Mr. ___ of "Peter Pan"

38. Fill in the blank with this word: ""The Aba ___ Honeymoon""

39. Water

41. Selected athlete

42. Poorly developed, as an idea

44. Strauss's "___ und Verkl‰rung"

45. The name of this one-celled protozoan comes from the Greek for "change"

46. Some shoes, for short

49. Soprano Nixon

50. Fill in the blank with this word: "Coeur d'___, Idaho"

51. Place to put the feet up

52. Many a holiday visitor / Bandit

53. Fill in the blank with this word: ""___-Man" (1974 spy/sci-fi film)"

54. Vitamin bottle info

55. Lot

58. New York's former ___ Building

59. Fill in the blank with this word: "___ Onassis, Jackie Kennedy's #2"

PUZZLE 38

ACROSS

1. Vanquished

5. Actress Samantha

10. Unmelodic sounds

14. Mythical king of the Huns

15. When one might have a late lunch

16. Bothers

17. Perfect plot

20. Persistence of memory concept

21. Plumbs the depths

22. Waters, informally

24. Part of an Asian capital's name

25. Fill in the blank with this word: ""Bali ___""

28. Texans' grp.

29. Either of two books of the Apocrypha: Abbr.

30. QB Detmer and others

33. About

35. Klinger player on "M*A*S*H"

37. Ready or not, here ___'

39. Hank Ketcham comic strip

42. Tony who led the N.L. in batting eight times

43. Unscramble this word: neam

44. Vardalos and Peeples

45. U.S. trading partner, formerly

46. The Missing Drink : High _____ rose

48. Title for one on the way to sainthood: Abbr.

50. What's right in front of U

51. Life of Pi' author ____ Martel

52. CPR pros

54. Frog sounds

57. West Flanders resort port

61. Chick lit book #1 (1992)

65. Fill in the blank with this word: "____ temperature"

66. Worrisome food contamination

67. Rare book dealer's abbr.

68. Kipling's "____ we forget!"

69. Like some tattooed characters

70. I. M. and Mario

DOWN

1. Restrain

2. People: Prefix

3. Without ____ to stand on

4. Un gato grande

5. Winter headgear

6. Ensured: Abbr.

7. WorldCom competitor

8. Grass part

9. Places to sleep

10. Heart of France

11. Wing: Abbr.

12. It may rock you to sleep

13. Princes, e.g.

18. Tiny battery

19. a person devoted to refined sensuous enjoyment (especially good food and drink)

23. Fill in the blank with this word: ""Everybody Jam!" singer ____ John"

24. Make grief-stricken

25. Waffle

26. Fill in the blank with this word: ""____ there yet?""

27. Fill in the blank with this word: ""Back ____" (1974 Genesis song)"

30. Surface-____

31. Workout spots, for some

32. Fill in the blank with this word: ""What thou ____, write": Revelation"

34. The Admiral Benbow ____ ("Treasure Island" locale)

36. Line score letters

38. This is ____'

40. The English translation for the french word: incorporation

41. of or relating to or involved the practice of aiding the memory

47. Fill in the blank with this word: "Eve ____, "The Vagina Monologues" monologist"

49. Wreck-checking org.

51. Vocally bother

53. New York ____

54. Fill in the blank with this word: "____-Alt-Del"

55. Actress Madlyn

56. 1936 Olympics hero

58. Fill in the blank with this word: ""____ kleine Nachtmusik""

59. 40's theater director James

60. R.A.F. awards

62. Smallest NATO member by population

63. Voting "nay"

64. Muhammad ____

PUZZLE 39

1	2	3	4	5	6		7	8	9	10		11	12	13
14							15					16		
17							18					19		
			20			21					22			
23	24	25				26					27			
28				29	30			31	32	33				
34				35			36		37			38	39	40
41			42				43							
44						45					46			
			47			48		49			50			
51	52	53				54	55				56			
57				58	59				60	61				
62				63					64			65	66	67
68				69					70					
71				72					73					

ACROSS

1. The English translation for the french word: tamtam

7. Fill in the blank with this word: "____-Ball"

11. The English translation for the french word: prier

14. Wool source

15. Woolen caps

16. Wore away

17. Small-plane maker

18. The sculptures "Cloud Shepherd" and "Coquille Crystals"

19. Fill in the blank with this word: ""Thanks for ___ Memory""

20. Connected on only one side, as a town house

23. Amazes

26. When some stores open

27. Fill in the blank with this word: "___ Southwest Grill (restaurant chain)"

28. either end of the yard of a square-rigged ship

31. Lux. neighbor

34. R&B singer with the hit 'It's All About Me'

35. Two-time N.B.A. M.V.P. Steve

37. Kind of rug

41. "Interview With the Vampire" actor

44. Promontories

45. Caught some rays

46. The ___-Mags (classic punk rock band)

47. Politico Hutchinson and others

49. Low-ethanol fuel blend

51. Oft-quoted auth.

54. Indian poet ___ Aurobindo

56. London's ___ Park

57. Good for a scientist, bad for a roofer

62. Fill in the blank with this word: "___ in apple"

63. Twin sister of Ares

64. Go from endangered to extinct

68. Year the emperor Decius was born

69. The English translation for the french word: rift

70. Crazy Horse, for one

71. Start for plop or plunk

72. Fill in the blank with this word: "___ one"

73. Yellowish-pink

DOWN

1. Washington State's Sea-___ Airport

2. Work without ___

3. Some sports cars

4. Like some salads

5. Obama education secretary ___ Duncan

6. Thank-you-___

7. Town: Ger

8. Best Supporting Actress nominee for "Five Easy Pieces"

9. Pre-___ (take the place of)

10. She, in Italy

11. Depth: Prefix

12. Fill in the blank with this word: ""With this ring ___ wed""

13. Exploits

21. Fill in the blank with this word: ""___ sorry!""

22. Mountain ridge

23. Adm. Rickover of the 40-Across

24. The 18th century's "Mad Anthony" or the 20th century's "Duke"

25. Sig Ep and others

29. Queen ___ War

30. Nancy's opposite number, once

32. Nine: Prefix

33. Hollywood's Alan and Diane

36. Quoits pegs

38. Don Marquis's six-legged poet

39. Title character of TV's "The Pretender"

40. Fill in the blank with this word: ""___ Mio""

42. With over 64 million in attendance, this Japanese city's 1970 World's Fair was the largest ever held

43. Prefix with -plasty or -gram

48. It's not automatic

50. "Rats!"

51. Defeat

52. Poet who wrote "I have executed a memorial longer lasting than bronze"

53. Race of Norse gods

55. B. & O. stop: Abbr.

58. The Last Time I Saw Paris' composer

59. Fill in the blank with this word: "Beethoven's "Archduke ___""

60. Some hairstyles

61. This prefix turns 1 byte of data into about a billion

65. When "S.N.L." wraps in N.Y.C.

66. Rapper ___ Rida

67. Zine reader

PUZZLE 40

1	2	3	4		5	6	7	8		9	10	11	12	
13					14					15				
16					17			18						
19				20								21		22
23			24				25				26			
27						28				29				
30					31				32					
			33	34				35						
36	37	38					39				40	41	42	43
44					45				46					
47				48				49						
50				51			52				53			
	54		55							56				
	57				58					59				
	60				61					62				

ACROSS

1. Fill in the blank with this word: "Air___, discount carrier"

5. Fill in the blank with this word: "Baseball's ___ Gaston"

9. Longtime Vicki Lawrence character

13. Fill in the blank with this word: ""___ Rebel" (1962 #1 hit)"

14. One way to race

15. Pile

16. I could ___ horse!'

17. capable of being shaped or bent or drawn out

19. Pie ___ mode

20. 1990s White House chief of staff Bowles

21. MacBook ___

23. Jeepers!' in Jersey

25. Fill in the blank with this word: ""My Name Is Asher ___""

26. "Goldberg Variations" composer

27. WordPerfect company

28. Street sign abbr.

29. Troy, in poetry

30. Watch part

31. Apt. feature, in the classifieds

32. Smooth

33. Coot

36. The narrator of „ÄúThe Curious Incident of the Dog in the Night-time‚Äù has this developmental condition

39. Santo Domingo greeting

40. Vermont but not New Hampshire, e.g.?

44. Johanna ____, author of "Heidi"

45. Upstate N.Y. school

46. Members of a raiding party

47. The sculptures "Cloud Shepherd" and "Coquille Crystals"

48. Tity ____ (Playaz Circle rapper)

49. Means of getting in touch

50. Reply facilitator: Abbr.

51. Think that might is right?

53. Ka ____ (Hawaii's South Cape)

54. "Whatever"

56. They're over specialists: Abbr.

57. Your majesty'

58. Wagering sites, for short

59. Stimulate, as the appetite

60. City in Judah

61. Unusual shoe spec

62. Near ____ South Dakota pageant

DOWN

1. Kindergarten recitation

2. Divide up in a new way

3. Star born Frederick Austerlitz

4. Sports org.

5. Toyota model

6. Certain supermarkets, for short

7. Fill in the blank with this word: "____ Day (September 19)"

8. More unctuous

9. Those, to Robert Burns

10. Neighbor of Syr.

11. Area of barren lava flows, in the Southwest

12. Quick impressions

18. Stationer's item: Abbr.

20. Fill in the blank with this word: "Electric ____"

22. Words of surprise

24. Some nonfiction works

26. Loud

28. Weed

29. QuÈbec's ____ d'OrlÈans

31. Teacher's deg.

32. LP successors

34. Long-running B'way musical seen by couples?

35. Ransom ____ Olds

36. Politico Hutchinson and others

37. Lift

38. Digitally entered

41. University in Garden City, N.Y.

42. Portrayer of Lt. Rodriguez on "N.Y.P.D. Blue"

43. Egg holders

45. Rod

46. Scoundrel

48. Was taken in

49. Fill in the blank with this word: "Bel ____"

51. Memory: Prefix

52. Player of Det. Eames on "Law & Order: Criminal Intent"

55. Fill in the blank with this word: "____ pro nobis"

56. Welsh valley

PUZZLE 41

ACROSS

1. Real estate ad abbr.

5. Turns the other way

9. Defeat

14. Fill in the blank with this word: ""Veni, vidi, ___""

15. Grand Ole ___

16. Gettysburg general

17. One with lots to think about

19. ___ You' (1975 #1 Minnie Riperton hit)

20. They're not given

21. Chemist's extracting solvent

22. Sportscaster Allen

23. Sound of a leak

24. aroused to impatience or anger

28. Where to see a mummy: Abbr.

29. Old Testament book

33. Slightly above average

34. Japanese writing form

35. Weeder's tool

36. Unquestioning adherents

40. Soprano Christiane ___-Pierre

41. Wildcats' org.

42. "Butter knife" of golf

43. Fill in the blank with this word: "___ perpetua (Idaho's motto)"

45. Red ___ (young amphibian)

46. Poorly developed, as an idea

47. "Billy Budd" captain

49. Tulsa sch. named for an evangelist

50. Tertiary Period part

53. Made a cashless transaction

58. Old post of the 7th Infantry Div.

59. Satisfactory

60. Guardian spirits

61. Field worker

62. Workplace fairness agcy

63. Weapon handles

64. Jazz saxophonist/flutist Frank

65. Ukr. and Lat., once

DOWN

1. Some underwear

2. Prie-___ (prayer bench)

3. Stereo syst. component

4. Personal air

5. Rocketed

6. Fencer's foot stamp

7. Tests for srs.

8. Six-Day War participant: Abbr.

9. Unloads

10. Common project in shop class

11. Thomas ___ Edison

12. Chili powder ingredient

13. Writer who coined the phrase "categorical imperative"

18. Suburb of San Diego

21. Writing by Montaigne

23. Noted jazz bandleader

24. Lonette of "The Cotton Club" and "Malcolm X"

25. Ubiquitous players

26. Soda jerk's creation

27. Suffix with pocket

28. Not talking to, perhaps

30. Valentino title role, with "the"

31. Yankees manager Joe

32. Surface anew

34. Facing the pitcher: 2 wds.

37. Possible answer to "Where are you?"

38. Quit your excuses'

39. Mauna ___ Observatory

44. No longer upset

46. "Good grief!"

48. The ___ near!'

49. Writer Joyce Carol ___

50. What I will follow

51. Fill in the blank with this word: "Cup ___ (hot drink, informally)"

52. Part of the N.C.A.A.'s purview: Abbr.

53. Lodge sign

54. Hydrocarbon suffixes

55. Wishes undone

56. Give an ___ effort

57. R.A.F. awards

59. Beat and how!

PUZZLE 42

ACROSS

1. Certain supermarkets, for short

5. Old capital of Romania

9. When some morning news programs begin

14. I never ___ man ...'

15. Word on a gift tag

16. This Houston energy company was delisted from the NYSE in 2002; later many of its execs pleaded the Fifth

17. Yield, as interest

18. Purchase from a jeweler

19. With 70-Down, do much (for)

20. Wily style of diplomacy?

23. Weaken

24. Taken ___

25. Regular: Abbr.

28. Sci. facilities

32. Sow bug or wood louse

34. Obama's signature health law, for short

37. Sporty car features

39. Sew up

40. With 36- and 58-Across, what the

answers to the starred clues are

44. Some alerts, for short

45. Wishes one can get on a PC?

46. One of the Chaplins

47. Toughish

50. Best boy's colleague

52. Some like it _____ plate

53. Wiring experts: Abbr.

55. Body of water in a volcanic crater, for one

59. Wee lad feted by the Friars?

64. Fill in the blank with this word: ""Coffee, ___ Me?""

66. Sam Shepard's "___ of the Mind"

67. Wyo. neighbor

68. Author Sinclair

69. Nein : German :: ___ : Russian

70. George Sand's "___ et lui"

71. Baltimore team, in sportspeak

72. Salinger's 'For ___ - With Love and Squalor'

73. Wet septet

DOWN

1. Marvin Gaye's '___ Little Girl'

2. Transmission

3. Venae cavae outlets

4. Holy, in Latin phrases

5. Then preceder

6. 1981 hit film with a 5'3" lead actor

7. Work with mail

8. Fill in the blank with this word: "___ ware (Japanese porcelain)"

9. "Blume in Love" star and others

10. Sinead O'Connor album 'Am ___ Your Girl?'

11. Novelty item in comic book ads

12. Yahoo! competitor

13. Telephone trio

21. Smallville family

22. Ransom ___ Olds

26. Fill in the blank with this word: "Cruiserweight boxing champion James ___"

27. Huey, Dewey, Louie, Donald and Daisy

29. Racing initials

30. Zipped

31. I Am Not ___' (1975 show business autobiography)

33. Every, to a pharmacist

34. Up to one's ears

35. Admit

36. Adjudicate

38. Hit a ___

41. U.S. mil. medal

42. Scottish Peace Nobelist John Boyd ___

43. The English translation for the french word: Udine

48. Pulitzer-winning historian Doris ___ Goodwin

49. Zorba imparts the beauty of the Greek language, but some lessons are tricky, like the one teaching that "ne" means this

51. Some fighters

54. Batting coach's concern

56. Give extreme unction to, old-style

57. Fill in the blank with this word: "Coleridge's "___ Khan""

58. ___ humains (people, in Paris)

60. Reset number

61. Golf innovator Callaway and bridge maven Culbertson

62. Per ___ (daily)

63. Tampa-St. ___

64. Pres. Hoover's dog King ___

65. N.T. book

PUZZLE 43

ACROSS

1. Largest employer in Newton, Iowa, until 2006

5. Where the last flight ends?

10. Org. whose workers may be left carrying the bag

14. Mahler's "Das Lied von der ___"

15. Fill in the blank with this word: ""___ Jacques" (children's song)"

16. Ashen

17. Lit ___ (college course, slangily)

18. Pope John Paul II's real first name

19. Sugar ___

20. [See circled letters]

23. Fill in the blank with this word: ""He's ___ nowhere man" (Beatles lyric)"

24. Tommy ___, Olympic skiing gold medalist

25. Slippery ___

28. These masses of fibrocartilage between vertebrae serve as shock absorbers

32. High-altitude home

34. On the Road' narrator ___ Paradise

37. See 43-Down

40. TV's "American ___"

42. Fill in the blank with this word: "Carry ___"

43. Obi accessory

44. Cafeteria

47. The English translation for the french word: lÈpisostÈe

48. Have a ___ stand on

49. Fill in the blank with this word: ""___ me!""

51. Visitors to the Enterprise

52. Fill in the blank with this word: ""Ready, ___...!""

55. Signs of healing

59. One who plunders boatloads of jack-o'-lanterns?

64. The English translation for the french word: pion

66. Of the ear

67. TV's 'How ___ Your Mother'

68. Ancient reveler's "whoopee!"

69. Sleep: Prefix

70. Hurdle for some univ. seniors

71. Workers need them: Abbr.

72. 1994 sci-fi epic

73. Short dogs, for short

DOWN

1. Pilgrim's destination

2. Fill in the blank with this word: "Computer ___"

3. Stops on ___

4. Start to fume

5. Injure

6. Fill in the blank with this word: "___ II (razor brand)"

7. Fill in the blank with this word: "___ Polo of "Meet the Fockers""

8. Fill in the blank with this word: ""...is sad and dreary ev'rywhere ___": Stephen Foster"

9. Yo-Yo Ma's instrument

10. Central knob of a shield

11. Making fun of mercilessly

12. Log cabin material, maybe

13. The Beatles' ___ Pepper

21. Lamb's "Essays of ___"

22. Unexciting marks

26. Arrowsmith's wife

27. Wayne ___ (Gotham City abode)

29. Hit a ___

30. Plasm prefix

31. Fill in the blank with this word: "___ Valley, Calif."

33. Fill in the blank with this word: ""Lord, is ___?": Matthew"

34. Walk laterally

35. On ___ (trying to lose)

36. Utensil for a 117-Across

38. Woe ___ them that call evil good': Isaiah

39. Unusual shoe spec

41. Kiddie ___

45. Linguist Chomsky

46. Spanish ayes

50. Save's partner

53. What a player may mean by knocking on the table

54. Nolan Ryan, notably

56. Transmission repair franchise

57. Substitute players

58. Small paving stones

60. Quelques-___ (some, in France)

61. Fill in the blank with this word: "___ la Douce"

62. -

63. Graceful bend

64. Zoologist's foot

65. They're stuck in I.C.U.'s

PUZZLE 44

ACROSS

1. Title for some bishops

5. About six centuries hence

9. "Fantastic" figure of children's lit

14. Louisiana ___: Abbr.

15. Seal's opening?

16. What's now in Mexico?

17. Kind of bike

18. Fill in the blank with this word: "___ of Attalos (Greek museum site)"

19. Works magic on

20. Dire proof-of-purchase slip?

23. Heart of France

24. Fill in the blank with this word: "Actor ___ Carroll"

25. ___ port

28. Officially listed: Abbr.

31. Aquamarine

33. Worrying sound to a balloonist

36. Tiny particle: Abbr.

38. Kol ___ (Yom Kippur prayer)

39. Me, in song

44. He can't hit the broad side of ___'

45. Stain stopper

46. Trains: Abbr.

47. Dweller in Pristina

50. Big Southern department store chain

53. Ordinal suffix

54. Word with code or road

56. Waits patiently

60. What a show horse may be in?

64. Texas cook-off dish

66. Give off, as fumes

67. The English translation for the french word: raclure

68. Stinky

69. Vittles

70. Presidential ___

71. Dissolves, as cells

72. U.K. military medals

73. Critic, at times

DOWN

1. Where the last flight ends?

2. Game of chance

3. Fill in the blank with this word: "Alla ___ (music notation)"

4. Hall-of-Fame golfer Tommy ___, called the "Silver Scot"

5. Like some nouns: Abbr.

6. Tiny bit

7. The English translation for the french word: baver

8. Irritates

9. Varieties of this tree, noted for its wood, include the West Indian & Honduras

10. Stat starter

11. Dana Scully's sci-fi partner

12. Silver ___

13. Fill in the blank with this word: "___ in xylophone"

21. Wrath

22. John ___

26. Rude and sullen

27. Kicks

29. Yukon S.U.V. maker

30. Fill in the blank with this word: "___ one-eighty"

32. Twice, in music

33. Wind

34. Wooden-soled shoe

35. Chart toppers

37. Overseas bar deg.

40. Mancinelli's "___ e Leandro"

41. Some shut-ins

42. Women's ___

43. U.K. honour

48. The American Recovery & Reinvestment ___ (A 3-letter word)

49. Stunk

51. Weight abbr.

52. Do-it-yourself wheels

55. Spanish counterparts of mlles.

57. "Beavis and Butt-Head" spinoff

58. The English translation for the french word: Ètude

59. When haroseth is eaten

61. Sam Shepard's "___ of the Mind"

62. Fill in the blank with this word: "___-1 ("Ghostbusters" vehicle)"

63. Sleep, in British slang

64. Toronto Argonauts' org.

65. You there!'

PUZZLE 45

1	2	3	4	5		6	7	8	9		10	11	12	13
14						15					16			
17					18						19			
20							21		22					
			23			24			25					
26	27	28	29		30			31	32					
33				34			35					36	37	38
39				40	41					42				
43				44					45					
		46					47	48		49				
50	51	52				53			54					
55				56	57		58			59	60	61	62	
63				64		65								
66				67					68					
69				70					71					

ACROSS

1. Bad-tasting

6. Word that can follow the start of 20-, 38- or 50-Across

10. Fill in the blank with this word: "1995 Physics Nobelist Martin L. ___"

14. Worth keeping, perhaps

15. Fill in the blank with this word: "___ podrida"

16. Poulenc's "Sonata for ___ and Piano"

17. Supermodel Evangelista

stuck in a dead end?

19. Workplace for Reps. and Dems.

20. Chat

21. Noted entertaining family from Utah

23. Part of a dict. entry

25. Wrapped up

26. Compass points (seen spelled out in 20-, 26-, 43- and 53-Across)

30. Barely cook, say

33. Difference in days

between the lunar and solar year

35. Second-century year

36. Film director ___ Lee

39. CAMERA!

43. Fill in the blank with this word: "___ agent"

44. White Sands Natl. Monument state

45. Fill in the blank with this word: "___ Zelnicek (celebrity's maiden name)"

46. Pretty good result for a golf round

49. Swedish actress Persson

50. You can count on them

53. Wedding exchange

55. Uses as support

58. Stuart Little' author

63. White house: Var.

64. Part of a blouse that touches the waist?

66. Fill in the blank with this word: "___-purpose"

67. Some nest eggs

68. Have another picture taken

69. Trix alternative?

70. Sentry's call

71. In April 1934 Mt. Washington had a wind gust reach 201 of this unit equal to 1 nautical mile per hour

DOWN

1. Fill in the blank with this word: "___ Morris, signature on the Declaration of Independence"

2. Fiery

3. Ravel's "Gaspard de la ___"

4. George Harrison's "___ It a Pity"

5. Pfeiffer of TV's 'Cybill'

6. Fill in the blank with this word: "Anne ___ (Henry VIII's second)"

7. Fill in the blank with this word: "Dryden's "___ for Love""

8. Nile queen, informally

9. Strikeout symbols, in baseball

10. Pennsylvania's ___ Mountains

11. U.S. investment instrument

12. Fill in the blank with this word: "1950 Max Ophuls film "La ___""

13. 1988 Peter Allen musical

18. Battle of ___ (1943 U.S./Japanese conflict)

22. Renowned family of Italian history

24. The annus in Dryden's "Annus Mirabilis"

26. 40's theater director James

27. What might be used with a 'Giddy-up!'

28. Rabbit ___

29. Kimono accessory

31. She, in S

32. Oysters ___ season

34. With 47-Down, title for this puzzle

36. Some small batteries

37. Ana

38. Fill in the blank with this word: "___ : hello :: hooroo : goodbye"

40. The Beatles' "___ Mine"

41. Uncle ___ of "Seinfeld"

42. Fill in the blank with this word: "1965 #1 hit "___ of Destruction""

46. Eightfold

47. One of the Crusader states

48. The ___ Report (upscale magazine)

50. Taken for ___

51. Under way

52. The lion in "The Lion, the Witch and the Wardrobe"

54. Preliminary drudgery

56. Fill in the blank with this word: "___ suspension (ear medication)"

57. Neighbor of Swed.

59. Coordinate in the game battleships

60. You never had ___ good!'

61. Hop ___!'

62. CPR pros

65. This word for a buddy or chum is often followed by "around"

PUZZLE 46

```
 1  2  3  4     5  6  7  8     9  10 11 12 13
14           15          16
17           18          19
20        21          22
23           24       25 26 27
28        29 30 31 32  33    34
      35          36     37
38 39 40          41 42
43        44
45        46 47    48       49 50 51
52        53    54      55 56
      57 58       59 60 61
62 63       64          65
66          67          68
69          70          71
```

ACROSS

1. The year 640

5. There have been 12 popes with this devout name, the first during the second century, the last from 1939 to 1958

9. Pork sandwich available periodically

14. Douay prophet

15. Fill in the blank with this word: ""King ___" (1950-65 comic strip)"

16. Fill in the blank with this word: ""I Didn't Slip, I Wasn't Pushed, ___" (1950 song)"

17. Dragsters' org.

18. Spur on a climbing iron

19. "___ disturb!" (and a hint for 17-, 40- and 65-Across, and 10- and 30-Down)

20. Cry of economic liberation?

23. Zane and Lady Jane

24. Toujours ___

25. The Unsers of Indy

28. Special ___

29. Some servers

33. With French, one of two official languages of Chad

35. Football squad

37. Greek goddess Athena ____

38. Local farmers

43. ____ Kalugin, former K.G.B. general with the 1994 book "Spymaster"

44. Went mad

45. On the other hand...'

48. Tear down, in England

49. War on Poverty agcy.

52. Fill in the blank with this word: "Bad ____, Mich. (seat of Huron County)"

53. Takeaway game

55. Fill in the blank with this word: ""What God has joined together, let ____ put asunder""

57. Meat-ban cause

62. Ecclesiastical assembly

64. The English translation for the french word: pion

65. Les …tats-____

66. Singer Jackson

67. High-tech transmission

68. Fill in the blank with this word: ""____ by me""

69. With 2-Down, 65 miles per hour, say

70. Snake sound

71. USMC rank

DOWN

1. Stay!'

2. The key of E major has one

3. Victor at Thermopylae, 480 B.C.

4. Like maples but not firs

5. Web ____

6. Latin hymn "Dies ____"

7. Tucson sch.

8. The first or fifth letter of George

9. Where trapeze artists meet

10. Corp. money managers

11. Like an attractive apartment

12. Worldwide workers' grp.

13. Triple-decker, perhaps

21. Where to sign a credit card, e.g.

22. Fill in the blank with this word: ""____ note to follow ...""

26. One who waits in ambush

27. Flies away

30. Food preservative letters

31. Work ____ many levels (succeed)

32. It's said to be salubrious

34. Texas ____ M

35. 1940s USSR secret police

36. Swift Malay boat

38. Fill in the blank with this word: "1998 French Open winner Carlos ____"

39. I'd like 'The New York Times Crossword' for $200, ____'

40. Take 4, clue 2

41. Woe while getting clean

42. Women's dress sizes

46. Like zombies

47. Fill in the blank with this word: "____-tac-toe"

49. Some Gulf Staters

50. Facilitating

51. Short court session?

54. Broods

56. Quiche ingr

58. Tom T. Hall's "Mama Bake ____"

59. Provider of a hot spot at a coffee shop?

60. Fill in the blank with this word: "____ told"

61. Band with the 1988 #1 hit "Need You Tonight"

62. U.K. distance measures

63. Whack

PUZZLE 47

ACROSS

1. Monk's hood

5. "Leave me alone!"

10. Naut. direction

14. Stub ___

15. "The Things We Do for Love" rock group, 1977

16. Fill in the blank with this word: ""King ___" (1950-65 comic strip)"

17. the doctrines and practices of the Presbyterian Church: based in Calvinism

20. Shoot-'em-up

21. Fill in the blank with this word: "Capitol-___ (music company)"

22. CafÈ cup

23. Taiping Rebellion general

24. Summed

27. Fill in the blank with this word: ""We've been ___!""

28. T-shirt size: Abbr.

29. Reply facilitator: Abbr.

31. 2002 Bond film

38. Takes a jog

39. Fill in the blank with this word: "___ Gonz"

40. Varsity QB, e.g.

42. Last: Abbr.

43. Leave high and dry

45. Reno and Kennedy, e.g.: Abbr.

46. Shed ___

48. Miles of film

49. Guardian spirits

51. Fill in the blank with this word: "___ to go"

53. Ratify anew

54. Third-place prize

58. Fill in the blank with this word: "Cole ___"

60. Using gold jewelry from the Israelites, he fashions the Golden Calf

61. They're held at both ends when eating

64. Fill in the blank with this word: "___ temperature (was feverish)"

65. Paint job finale

66. Explosives

67. Wine taster's adjective

68. Excelled

69. Wrapped (up)

DOWN

1. Horse of the Year that won the 1949 Preakness and Belmont

2. Others: Sp.

3. Fill in the blank with this word: ""___ him who believes in nothing": Victor Hugo"

4. Fill in the blank with this word: "___-majestÈ"

5. Suffix with theatrical

6. Rationed (out)

7. One of the three original Muses

8. The English translation for the french word: mordant

9. Year in St. Gregory I's papacy

10. Yemen's capital

11. Actress Van Devere

12. Fill in the blank with this word: "___ nova"

13. Like the Pantheon

18. Sonia of 'Kiss of the Spider Woman'

19. Stock phrase

25. With Altair and Vega, it forms the Summer Triangle

26. Fill in the blank with this word: ""I swear I ___ art at all": "Hamlet""

28. Uncle ___ of "Seinfeld"

30. Teacher's deg.

31. Mississippi's ___ State University

32. Like wire transfers

33. Speed-skating champ Johann ___ Koss

34. Utensil's end

35. Once I ___ secret love...'

36. 1969 Frank Sinatra album featuring Rod McKuen songs

37. Mystics

38. Fill in the blank with this word: "___ gum"

41. Some CBS forensic spinoffs

43. The Ponte Vecchio crosses it

44. See 1-Across

47. Windpipe, e.g.

50. Periods added to harmonize the lunar and solar calendars

52. Flock : birds :: cloud : ___

53. French income

55. Fill in the blank with this word: "">Scrubs" co-star ___ Braff"

56. Switch suffix

57. Sound stressed, maybe

58. With 17-Down, a temporary urban home

59. Fill in the blank with this word: ""___ note to follow ...""

62. Start of an aside, to tweeters

63. W-2 info: Abbr.

PUZZLE 48

1	2	3	4	5		6	7	8	9		10	11	12	13
14						15					16			
17						18					19			
20					21					22		23		
			24					25		26				
27	28	29	30				31	32						
33					34	35								
36			37		38				39		40	41	42	
		43		44			45		46					
47	48	49						50	51					
52						53								
54			55		56	57				58	59	60	61	
62		63	64					65						
66			67					68						
69			70					71						

ACROSS

1. Joe Jackson's "___ Really Going Out With Him?"

6. Fill in the blank with this word: "___ told"

10. Goatee, e.g.

14. Finger

15. Teatro ___ Scala

16. Fill in the blank with this word: "___ Fjord"

17. Silver coins of ancient Greece

18. Outfielder Mondesi

19. Trick ending?

20. Short or long, in phonetics

23. Phone no. add-on

24. Verdi's "___ tu"

25. You got that right!'

27. Zipper alternative

31. Small, slim daggers

33. S. American land

34. Some playful pups

36. The English translation

for the french word: akita inu

38. Fill in the blank with this word: ""___ Blue?" (1929 #1 hit)"

39. Workout spots, for some

43. Didn't just follow around

46. Fill in the blank with this word: "___ Aarnio, innovative furniture designer"

47. NyQuil ingredient?

50. Went by dugout

52. Jailed, slangily

53. TV/___

54. Sports org.

55. Unplanned

62. Requests for developers: Abbr.

64. Fill in the blank with this word: "___ Polo of "Meet the Fockers""

65. Songwriter Carmichael

66. They say this lady will ‚Äúcoax the blues right out of the horn‚Äù & ‚Äúcharm the husk right off of the corn‚Äù

67. Fill in the blank with this word: ""___ your pardon?""

68. Fill in the blank with this word: ""This ___" (shipping label)"

69. Fill in the blank with this word: "Alumni ___: Abbr."

70. Crepe paper feature

71. Popular

DOWN

1. Spontaneous skits

2. Fill in the blank with this word: "Chris ___, 1988 N.L. Rookie of the Year"

3. Warehouse

4. Swimming ___

5. Fill in the blank with this word: "Eve ___, "The Vagina Monologues" monologist"

6. Fill in the blank with this word: ""___ ask ...""

7. Mechanic's ___

8. Fill in the blank with this word: "Chug-___"

9. Square cracker

10. Talking-___ (scoldings)

11. Virgin Isl., e.g.

12. Like some muscles

13. Rich pastries

21. Scientology's ___ Hubbard

22. Like the Wolfman

26. Some meter readers

27. Miles of film

28. Theologian who opposed Martin Luther

29. Moli√°re's 'Le M√©decin Malgr√© ___'

30. Watch a kitty

31. Ice cream drinks

32. Sheriff Taylor's son

35. Fill in the blank with this word: ""___ show you!""

37. Where the last flight ends?

40. Wharton grad's aspiration, maybe

41. Fill in the blank with this word: ""Chances ___," 1957 #1 hit"

42. Turf

44. Pole position?

45. 800, to Caesar

47. The English translation for the french word: cinÈma

48. large sweet fleshy tropical fruit with a terminal tuft of stiff leaves

49. Writings by David

51. Vaulted

53. Winston Churchill flashed it

56. Semitic lang.

57. Whose woods these ___ think...': Frost

58. Work on the edge?

59. Wham-bam-thank-you-ma'am types

60. Tropical fever

61. Fill in the blank with this word: ""___ Blue""

63. Fill in the blank with this word: "China's Sun Yat-___"

PUZZLE 49

ACROSS

1. Tiddlywink, e.g.

5. Subjective pieces

10. Sodium hydroxide, to chemists

14. San ___, Italy

15. Singer-songwriter Jones

16. Penny-___ (trivial)

17. Fill in the blank with this word: ""___ (So Far Away)" (1982 hit by A Flock of Seagulls)"

18. 1962 war epic, with "The"

20. Miami's ___ Bay

22. Wharves

23. Italian emporium ending

24. France's Cote d'___

26. Reaction from one who has a bone to pick?

29. Mythological hunter turned into a stag and killed by his own dogs

32. London lockups

33. Red as ___

34. Lana Del ___, singer with the 2014 #1 album 'Ultraviolence'

36. Landers and others

37. Longtime Yes drummer

38. The English translation for the french word: saisir

39. Rock's ____ Soundsystem

40. British runner Steve

41. Wake Island, for one

42. Tractor-drawn fall activity

44. Playwright Sean

45. Cub #21 of the 1990s-2000s

46. Like an N.B.A. team

47. Winningest southpaw in major-league history

50. Skateboard wheel material

54. Where to belt one down and belt one out

57. Massachusetts' ____ College

58. Fill in the blank with this word: ""Which Way ____?" (1977 film)"

59. Title for Sulu on "Star Trek": Abbr.

60. Walk back and forth

61. You can get one on the house: Abbr.

62. Std. on food labels

63. Squealed cries

DOWN

1. Wee bit

2. Yesterday, in Italy

3. Shirt sizes

4. Masks

5. Not unless

6. City ESE of Bombay

7. White-tailed eagle

8. The U.N.'s ____ Hammarskj

9. Fill in the blank with this word: "____-wolf"

10. Unscramble this word: tnuaer

11. Time ____ half

12. Buckwheat's affirmative

13. Attention-getters

19. Start of a weightlifting maneuver

21. Some airport data: Abbr.

24. Not worth ____

25. Beginning on: 2 wds.

26. Fill in the blank with this word: ""I've Got ____ in Kalamazoo""

27. Where you're likely to see dirty hands

28. "The Grapes of Wrath" star, 1940

29. Weaken

30. Two-tone treats

31. Writer Zora ____ Hurston

33. Tree-lined promenade

35. Every 12 mos

37. Rara ____

38. Light, one-seated carriage

40. Waste

41. Room to swing ____

43. A wishbone has one

44. Mario Puzo best seller

46. Whistle-blower's exposure

47. Whole alternative

48. Yesteryear

49. Fill in the blank with this word: "Cut ____ "

50. Fill in the blank with this word: ""Deutschland ____ Alles""

51. Zoological wings

52. Small cut

53. Hydrocarbon suffixes

55. Nickname of 1954 home run leader Ted

56. Sue Grafton's '____ for Evidence'

PUZZLE 50

1	2	3	4		5	6	7	8	9		10	11	12	13
14					15						16			
17					18					19				
20				21			22					23		
			24		25	26					27			
28	29	30						31						
32				33			34				35	36	37	38
39			40					41	42	43				
44						45					46			
			47	48	49	50		51			52			
53	54	55				56								
57				58					59		60	61	62	
63			64				65	66			67			
68				69							70			
71				72							73			

ACROSS

1. Grp. organizing '60s sit-ins

5. These nomadic people of southern Kenya build huts in a circle to create a boma, or village

10. Pack ___ (quit)

14. City of northern Finland

15. Suffixes with depend and differ

16. Oxford bottom

17. Mr. ___ of "Peter Pan"

18. an acquaintance that you go to school with

20. Possible answers for 20-Across

22. Pas ___ (gentle ballet step)

23. Massenet's "Le ___"

24. Saying of Caesar

28. Part of a jazz combo

31. Zaire's Mobutu ___ Seko

32. Untilled tract

33. Uninteresting

35. Solidarity's Walesa

39. Art form in Quebec?

44. Sound of a leak

45. Time for eggnog

46. Trickiness

47. Ticket sellers: Abbr.

51. Reproducible in great detail

53. Star of TV's "Veronica's Closet"

57. One in the charge of un instituteur

58. Suffixes with ballad and command

59. Visionaries

63. Smooth-skinned fruit

67. See 83-Down

68. Skin Bracer alternative

69. Brutus's burdens

70. The Neverending Story' author

71. Was ___ hard on them?'

72. Wet, in a way

73. Fill in the blank with this word: "___ Trueheart of "Dick Tracy""

DOWN

1. Cub #21 of the 1990s-2000s

2. Unfeeling

3. Nile queen, informally

4. Tequila brand, for short

5. Fill in the blank with this word: "___ amis"

6. Mandela's org.

7. Cold sorrel soup

8. ancient name for the coastal region of northwestern Asia Minor (including Lesbos)

9. In medieval legend: Tristan & ____

10. Fill in the blank with this word: ""___ Woman" (1972 #1 song)"

11. One of the singing Braxton sisters

12. Suffix with material

13. Fill in the blank with this word: ""___ say more?""

19. Visitors learn how to make these floral necklaces at Senator Fong's plantation & gardens

21. French film award

25. This British term for a police spy or informer sounds like it's related to drugs, but it dates from the mid-19th C.

26. Showy flower of the iris family

27. South African grassland

28. Whittier poem "___ Well"

29. Wings: Lat

30. Shooters

34. Letters after Daniel Moynihan's name

36. White House's ___ Room

37. Year Attila was defeated in Gaul

38. Hic, ___, hoc

40. Those: Sp.

41. The Shelters of Stone' writer

42. Mae West role

43. Yellow-fever mosquito

48. Lack of vigor

49. Steering system part

50. Amino acid in many proteins

52. Shoe part

53. Alaskan peninsula where Seward is located

54. Irving Berlin's "___ My Heart at the Stage Door Canteen"

55. Verso's opposite

56. Good ___ (repaired)

60. Fill in the blank with this word: ""___ kleine Nachtmusik""

61. Some mail designations: Abbr.

62. Women with aur

64. Way of the East

65. We'll teach you to drink deep ___ you depart': Hamlet

66. Woeful

Solutions

Puzzle Solution 1

C	A	S	H	■	A	S	S	I	S	I	■	C	P	L
A	M	P	S	■	L	E	C	H	E	S	■	H	E	E
P	A	Y	T	H	E	P	I	P	E	R	■	I	E	R
O	H	H	■	E	A	T	■	■	A	A	L	T	O	■
■	■	■	E	L	V	I	S	P	R	E	S	L	E	Y
L	E	N	N	I	E	■	P	I	E	L	S	■	■	■
A	L	I	T	O	■	E	L	L	I	O	T	T	S	■
H	E	B	R	■	B	C	E	L	L	■	R	O	R	T
R	E	S	E	A	L	E	D	■	■	O	T	W	A	Y
■	■	A	N	I	C	E	■	A	V	E	N	G	E	■
L	I	F	T	O	P	E	R	A	T	O	R	■	■	■
L	L	O	Y	D	■	■	T	P	I	■	W	H	Y	■
O	E	R	■	I	T	C	O	U	L	D	N	T	B	E
S	A	M	■	Z	A	D	O	R	A	■	S	W	A	G
A	C	A	■	E	D	P	O	N	Y	■	F	O	G	G

Puzzle Solution 2

R	O	P	Y	■	L	I	A	M	■	P	E	L	L	A
E	L	L	E	■	E	T	U	I	■	U	N	S	A	Y
C	L	A	S	S	W	O	R	K	■	T	S	I	D	E
E	A	G	L	E	D	■	A	E	R	I	■	■	■	■
■	S	E	E	D	E	R	■	D	O	T	D	A	S	H
■	■	T	U	R	N	S	■	A	T	M	F	E	E	■
D	E	I	S	M	■	A	P	P	R	O	V	A	L	S
I	T	S	■	I	S	A	A	K	■	■	L	A	H	■
R	H	I	N	O	C	E	R	I	■	F	A	L	S	E
G	Y	R	A	T	E	■	E	R	R	O	N	■	■	■
E	L	A	N	T	R	A	■	S	O	C	I	A	L	■
■	■	O	S	I	P	■	N	U	M	B	E	R	■	■
A	L	B	O	M	■	L	A	C	E	S	I	N	T	O
P	O	L	K	A	■	E	C	O	L	■	S	E	M	E
T	I	T	A	N	■	S	K	Y	Y	■	M	R	E	S

Puzzle Solution 3

```
M A C Y █ M I L D █ H O N K
E I R E █ I D I O M █ D K N Y
E D E N █ D I E G O █ T A E L
C E N T E R O F G R A V Y █ █
E R N A N I █ █ Y A D █ B S A
S S A █ C F C S █ L I B Y A N
█ █ J E F F E R S O N M A N █
S E E A █ O R U █ █ E E R S █
I R F I N E S T H O U R █ █ █
R I F L E D █ S R A S █ T E M
E C U █ R I A █ K F S E L C █
█ S O F T S H E L L C L A M █
M A I M █ H O O H A █ A L I I
A T V S █ S O R E N █ L E N I
A L E K █ F O L D █ A R E I █
```

Puzzle Solution 4

```
Z T E W █ M I D G E █ C P U S
A I R H █ I N R U N █ R A L F
R E N I █ N C A A S █ A S E C
F R O Z E N A C C O U N T S █
█ █ B R I S █ U Z I █ █ █ █
P A V A N E █ R E L I A N C E
C H A N S █ M U L E S █ A N N
K O N G █ T U N E D █ P H O S
L O N █ C H I T A █ B L U T O
E T A I L E R S █ F R A M E R
█ █ N E N █ S L A Y █ █ █ █
A I M F O R T H E S T A R S █
M C D I █ R O H A N █ I D L S
I T A N █ T T O P S █ M O C S
F A H D █ H I R E E █ E G G S
```

Puzzle Solution 5

```
A N O S E   R A N D R   S U P
T E N C C   O H I O U   H C E
Y O U R H O P E S U P   O L E
A N S A   S E N   B E D P A N
    W E S     S T R E W
  B A L L I S O N S T R I K E
S E M   M A O R I     E N N A
P A A V O   R A T   E N D I S
A T L I   E N C L S   O T E
D I G E S T S T H E N E W S
  A W H I T   C E R
A I M S A T   O R A   S O I E
L A A   P L O W E R P A N T S
L S T   E E L E D   O T T I S
E K E   D D D D D   O Z O N E
```

Puzzle Solution 6

```
A S L   I R A N T   E D A T E
S H E   N O S E R   N E V I N
T E N   T Y C H O   B W A R D
O L D F A S H I O N E D
P F L U G       P Y R R H I C
    R L E S S   E G O I S T
M T T   I L L I N   P F U I
A I R T O A I R M I S S I L E
U T A H   T E E N Y   S I S
R E C E S S   E X A M S
A R I E T T A     M E R I T
  N E E S O N N E P H E W
M E N D A   A T E A T   E R I
A R O O M   H O L B R   T E L
S E R F S   I S L E Y   T I L
```

Puzzle Solution 7

A	C	A	S	T		M	A	S	C		S	Q	I	N
S	O	R	E	H		A	L	A	I		N	O	N	A
D	R	U	P	E		L	I	N	T		U	P	O	N
F	A	N	T	A	S	T	I	C	E	I	G	H	T	
		A	X	H			T	E	N	G				
V	E	A	L		A	L	L	I		A	L	L	O	K
I	N	D		A	M	I	E		A	F	E	A	R	D
S	A	D	O	M	A	S	O	C	H	I	S	T	I	C
A	T	A	R	U	N		X	B	O	X		I	N	F
S	E	P	A	L		M	I	C	R		I	N	G	E
		N	E	U	E			S	A	S				
	B	E	G	T	H	E	Q	U	E	S	T	I	O	N
T	O	R	I		U	K	E	S		I	L	E	D	E
I	T	I	S		R	E	I	N		G	E	A	R	B
U	H	U	H		A	R	I	A		N	S	T	A	R

Puzzle Solution 8

M	D	D	M		M	L	I	I	I		O	N	O	
I	C	E	I	N		O	O	N	A	S		V	O	P
L	O	U	C	O	S	T	E	L	L	O		E	V	E
O	N	T	A	S	K			A	S	P	I	R	I	N
		S	E	I	N	E		O	O	M	P	A	H	
M	B	A		J	A	I	L	S		D	M	A		
E	E	W		O	R	E	O	O	S		I	S	A	W
S	T	E	U	B	E	N		P	A	R	E	S	I	S
A	S	S	N		A	T	A	S	T	E		E	D	M
		C	L	I		E	T	U	I	S		D	A	S
S	T	R	A	D	A		O	P	N	I	N			
D	R	A	Y	A	G	E			O	Z	O	N	I	C
L	U	V		H	E	Y	N	I	N	E	T	E	E	N
T	R	E		O	N	R	I	O		R	I	N	S	E
G	O	N		S	A	E	N	S		F	E	T	T	

Puzzle Solution 9

N	A	M	E			I	Y	W	M		R	A	N	A	C
A	C	T	V		Q	U	O	I		E	L	O	R	O
N	O	W	I		T	A	L	L		C	A	R	O	N
K	I	T	C	H	E	N	F	L	O	O	R	I	N	G
I	N	F	T	H	S		I	I	I	I				
			O	T	B	S		L	L	O	Y	D	S
M	A	T	S	U		C	H	E	E		R	O	M	Y
E	T	H	E	R	I	C		M	D	I	G	R	A	S
E	T	A	L		L	I	E	U		I	S	E	N	T
K	A	R	Z	A	I		H	S	N	I
			T	A	H	E		A	I	D	E	R	S
B	L	A	N	K	C	A	R	T	R	I	D	G	E	S
S	O	D	O	I		B	E	O	K		D	A	I	S
U	S	E	O	N		I	A	L	E		D	L	C	S
P	E	E	P	S		B	L	E	D		D	E	E	S

Puzzle Solution 10

A	W	A	C	S		K	G	S		A	G	R	E	E
J	A	Z	Z	Y		H	R	E		A	A	H	E	D
A	D	E	E	R		R	E	N	T	A	R	O	O	M
M	E	R	C	I		D	N	A	L	A	B
		B	H	A		N	A	R	C		A	L	E	G
P	S	A	S		I	N	D	Y		N	O	L	L
E	C	I		R	R	N	A		M	A	Z	U	M	A
P	A	J	A	M	A	S		D	E	P	O	N	E	D
C	L	A	U	S	E		A	I	D	A		G	R	E
I	A	N	S		M	M	S	U		C	E	S	S
D	R	I	P		E	A	R	H		A	L	S
		I	C	E	C	A	P		N	E	H	I	S
A	I	R	C	A	N	A	D	A		A	V	A	N	T
A	R	R	E	T		W	I	N		R	E	R	A	N
M	A	R	S	H		S	O	S		T	R	E	S	S

Puzzle Solution 11

I	L	O			G	A	T			U	P	L	O	A	D	
S	I	N			U	M	W		N	E	W	C	R	O	P	
A	M	E	R	I	C	A	N	I	N	P	A	R	I	S		
A	B	A	O	N			U	T	A			O	N	S		
C	O	M	M	E	R	C	I	A	L	C	O	W				
			Y	A	C	H	T	S			M	A	H	O	N	
S	L	S			P	E	S		P	O	T	E	N	T		
S	I	L	E	N	T	R		S	U	N	H	A	T	S		
B	R	A	D	Y	S		K	I	L			D	O	B		
Y	A	L	E	S		T	I	T	L	E	S					
		O	N	E	T	H	O	U	S	A	N	D	A	D		
G	A	M			Y	E	W			T	I	M	L	Y		
A	L	I	E	N	C	R	A	F	T	S	P	A	C	E		
P	A	N	A	C	H	E		S	O	U		S	E	R		
	I	G	L	O	O	S		T	W	P		S	E	S		

Puzzle Solution 12

T	H	A	I		L	A	P	T		G	I	G	U	E
B	A	L	M		A	M	S	O		R	E	A	R	S
S	T	I	N	G	R	A	Y	S		A	R	M	A	S
P	I	T	O	N		L	C	E	S	Z		U	W	E
S	N	O		A	F	G	H	A	N	I	S	T	A	N
		A	R	R	A	U		E	E	G				
H	E	A	T	L	A	M	P			T	T	O	P	
U	N	R	E	S	T	S		A	S	A	M	U	L	E
A	T	C	O			E	M	A	N	A	T	E	D	
	U	A	L		M	A	R	Y	J					
P	L	A	T	T	S	B	U	R	G	H		O	M	I
N	O	L		K	U	A	L	A		O	T	R	A	S
S	O	U	C	I		S	A	N	D	W	E	D	G	E
U	N	M	A	N		E	T	T	E		N	E	U	R
P	S	S	T	S		L	E	H	I		A	R	S	E

Puzzle Solution 13

```
W H I G   O A F S     T O R A
H O A R   E M A L L   L U A S
U R S I   J O N A Y   C T N S
P O K E R P L A Y E R   S D I
    S L E E T     L I L I A N
S T R E S S   I S L A N D
M R E   C O C A   T E E H S
T O A H A I R   E R A S M U S
E D G E D   S O S O     A R E
    A N N I K A   A M E N D E
H O N D A S   T I D A L
E H S   N E W B O R N B A B Y
A I R H E A R N A   O B I E
R O A R   A R A I C   W A R D
N U Y S   E N C E   S T L O
```

Puzzle Solution 14

```
B I O N   P S A S     R U H R
C R I B   A A B A   L A P A T
D I S C O N T E N T I S T H E
E D E   R E A D   M E S O N S
    O I L Y   V I O L
D I T Z E S   M I N N E L L I
E R V I N   H U T U   E E N
G I V E T H E B U S I N E S S
U N C   O C A S   B I D E T
M A R I A C H I   M E S S R S
    M I K E   L I R I
A R S E N E   Y O G I   M S S
G U I L T Y A S C H A R G E D
R E D D I   P E A T   Y M H A
A R E A   U R L S   S T R K
```

Puzzle Solution 15

A	T	W	T		S	T	E	N			T	O	P	U
B	E	E	P		P	A	S	H	A		I	D	L	S
M	A	D	I	S	O	N	I	L	L	I	N	O	I	S
S	C	S		W	R	A	T		F	T	E	R	E	R
		E	A	T			M	A	M	A				
T	H	I	N	G	S	T	H	A	T	B	R	E	A	K
O	E	S	T	E		A	U	R	A	E		N	S	A
S	N	O	R		S	T	R	O	H		L	O	O	P
I	C	U		B	A	L	O	O		P	U	T	U	P
T	H	R	E	E	P	E	N	N	Y	O	P	E	R	A
		P	A	I	R			O	L	E				
W	A	S	H	M	E		S	I	K	A		P	A	R
Y	O	U	R	E	N	O	T	S	E	R	I	O	U	S
E	U	R	O		T	W	I	L	L		D	E	B	T
S	T	A	N		S	R	A	S		E	M	E	U	

Puzzle Solution 16

B	G	A	M	E		D	N	G		U	S	R	D	A
U	R	B	A	N		I	B	A		G	H	E	N	T
N	O	O	I	L		A	A	R		L	O	M	A	N
C	A	R	D	S	E	L	E	C	T	I	O	N		
O	T	T	S		S	T	R	I	P	E		A	N	T
S	S	S		I	S	O		A	K	R	O	N	O	H
		E	V	E	N	I				B	T	W	O	
T	H	E	L	A	N	E	O	N	E	W	A	S	I	N
S	O	R	B				O	F	M	A	N			
P	R	I	E	S	T	S		L	O	L		D	A	M
S	A	C		E	A	T	S	A	T		N	O	W	A
		S	A	T	U	R	A	T	E	D	F	R	A	T
Y	U	S	E	F		A	R	I		I	L	I	K	E
O	N	O	N	E		T	E	O		D	E	T	E	R
U	L	N	A	E		O	E	N		I	R	O	N	S

Puzzle Solution 17

```
S T L O . . . A G S . O P E S
O R A D . H A D A T . S A M M
F I R E E S C A P E . A G U A
T V D . L E E R I N G . I L L
L I E W I T H . N O M I N A L
Y A R D S . I N G . A B A T E
. . S T A G S . C N O T E S .
S A M . W H Y M E . E S T . .
A D A P T S . N A D E R . . .
G A R D E . A C E . Q A N D A
U M M Q A S R . S A U R I A N
A B O . S U M A T R A . C M I
R E S H . S P I R A L P A S S
O D E S . H I T I T . E E E E
S E T I . I T S . G A L S . .
```

Puzzle Solution 18

```
A L L O R . V I O L . I M A M
B R A V E . E S N E . V O L E
M O D E L T R A I N G A U G E
S N E R . R I C O . O N E A R
. . . A S I T . N A O H . . .
T H E G A M E I S A F O O T .
R O P E R . M E H . E C O L .
E S S . I Z Z A T S O . U T A
O T O E . T I M . E M L E N .
. A M S T E L S T A N D I N G
. . C O W L . I T S I . . . .
R E D U B . I A T E . G R O G
L E A D E R O F A M E R I C A
C E R O . A N O N . S A G A L
G E E S . S S R S . P S S S T
```

Puzzle Solution 19

K	N	O	B		I	T	A	N			I	H	O	P
H	O	R	O		N	O	R	M		A	M	E	B	A
B	L	O	W	G	A	B	R	I	E	L	B	L	O	W
D	A	S		A	B	E	S		M	A	I	D	E	N
			S	L	A	Y		S	E	R	B			
P	S	C	H	E	R		R	E	E	M	E	R	G	E
A	N	C	O	N		P	A	R	R			A	S	L
L	O	V	E	A	T	F	I	R	S	T	B	I	T	E
M	O	I			E	L	S	A		E	S	T	A	N
E	P	I	L	O	G	U	E		A	N	I	T	R	A
			C	O	N	G		R	O	U	X			
R	E	S	H	O	E		B	E	L	T		A	A	M
E	Q	U	A	L	R	A	Y	L	E	O	N	A	R	D
C	U	R	I	A		B	E	O	R		A	B	E	L
S	I	A	M		M	S	G	S		H	A	D	I	

Puzzle Solution 20

L	A	P	P		O	N	E	I		T	R	I	M	
E	N	Y	O		G	A	R	D	E		R	E	C	U
I	N	R	O	A	D	S	I	N	R	H	O	D	E	S
S	A	O		B	E	O	N		R	A	I	D	E	D
		U	R	N			C	A	S	K				
I	S	T	H	I	S	S	E	A	T	A	K	E	N	
G	H	O	U	L		E	N	N	I	O		I	L	O
A	R	M	H		M	E	D	O	C		S	N	I	P
V	E	E		S	U	S	I	E		S	E	T	S	A
E	D	I	B	L	E	U	N	D	E	R	W	E	A	R
			H	O	S	P		L	E	S				
A	S	C	O	O	L		A	L	A	P		X	D	B
C	H	I	P	P	I	N	G	S	P	A	R	R	O	W
A	U	R	A		X	O	U	T	S		Y	V	M	Q
D	E	C	L		L	E	S	E		A	S	O	K	

Puzzle Solution 21

```
S T I P E ■ T D P ■ Y A K O V
T O N E S ■ R E A ■ O M N I A
A D A P T ■ A C R ■ G O E S T
L A D Y O F L E I S U R E ■
I T A S ■ L A M A R R ■ B E E
N E Y ■ C U L ■ H O T D O G S
■ U L N A E ■ A N E S
T H E B U G S T O P S H E R E
A B A R ■ D U A L S ■
R O S A L I A ■ T I S ■ S B A
T S Y ■ U N R O O F ■ O L E S
■ P E T T Y O F F I C E R S
L I O N H ■ A N U ■ P H A G E
P R U N E ■ N A S ■ O O Z E S
N A R E R ■ S S E ■ S A E N S
```

Puzzle Solution 22

```
T E R N ■ C O W ■ E B L A
I L A Y ■ O P I N E ■ L A O S
G U T T E R B A L L ■ I N N S
R A I ■ C A B O O S E ■ D E E
I N T R A D E ■ A H B L I S S
S T E P S ■ L E N ■ R O T O S
■ T H O L E ■ Y O Y O M A
E B B ■ B Y L O T ■ S E S
S A L A M I ■ E N D A T
T H I S I ■ C D P ■ F A T S O
E R N E S T O ■ A E R I A L S
E A D ■ O W N U P T O ■ B A I
M I E N ■ I R R E R S I B L E
E N R Y ■ S A B R E ■ B E O R
D I S S ■ T D S ■ A D M S
```

Puzzle Solution 23

N	A	N	O	S		S	L	O	A	N		S	H	E
I	D	O	N	T		O	E	S	T	E		A	A	M
B	O	O	S	T	E	R	I	S	M	S		L	S	E
S	T	R	E	E	T		E	S	T	H	E	T	E	
			T	R	U	A	N	T		L	A	S	E	R
C	A	L		E	N	D	O		S	E	A	W		
A	M	I	R	S		D	I	P	L		G	O	Y	A
V	A	N	U	A	T	U		A	R	T	E	M	U	S
S	H	E	M		S	P	Q	R		U	N	E	R	P
	B	O	O	P		U	S	S	R		N	I	S	
C	H	A	U	D		S	O	I	G	N	E			
L	U	C	R	E	C	E		N	O	M	E	A	T	
A	L	K		T	O	P	A	Z	Q	U	A	R	T	Z
S	A	E		T	E	T	R	A		T	I	N	E	A
S	S	R		E	N	I	A	C		S	L	I	E	R

Puzzle Solution 24

C	A	M	P		A	D	Z	E	S		E	C	G	S
O	N	U	E		S	E	I	K	O		N	A	O	H
F	O	L	L	O	W	I	N	G	O	R	D	E	R	S
F	I	L	E	N	A	M	E	S		A	I	S		
E	N	A		A	S	O		R	I	T	U	A	L	
E	T	H	N		S	M	L	E	D		R	T	E	
			O	C	A		U	L	C	E	R	A	T	E
	B	E	A	S	T	O	F	B	U	R	D	E	N	
R	O	Y	A	L	I	S	T		R	S	A			
A	N	E		E	O	S	I	N		S	O	I	L	
H	S	D	O	W	N		I	S	P		C	M	A	
		R	A	I		M	A	T	C	H	P	L	A	Y
E	R	O	S	S	A	W	I	W	A	S	S	O	R	E
R	E	P	T		S	A	W	I	N		A	C	E	R
S	I	S	S		C	H	A	T	S		T	K	T	S

Puzzle Solution 25

```
D E E R   G H A T   ■ B E E B E
E N N E   O O L A   E N R O N
G O O D W I L L B U N T I N G
A R C S I N E   O H I O A N S
S E H E N     F R O G
      A D A M A   H N D T Y Q
S S T   U M E K I   C H O U
T H E A P P L E O F O N E S I
K H A N     O S T E R   M T N
S H L E P P   I A M S O
      R E G T     I P A N A
D O G S E A R   M E N T E E S
W H A T S H O U L D I S T A Y
A N N E E   W E V E   I N R E
N E T W T   L Y I N   N A S T
```

Puzzle Solution 26

```
M D C C   V I C I   R O G G E
A I R H   E V A C   H A O L E
I D E A   R E R E   U H L A N
M I S S I S S I P P M U D
    M I I   B O O B   E M E
P A S S I O N   P L A I N T S
O F T   I N I S     P R O T
B R I T I S H A L E H O U S E
B A C H     G E N A   L S E
L I K E N E W   N F L T E A M
E D S   T I E D   R A I
    I D E N T I C A L T W I N
A R T E S   T A S M   I H R E
L E T A T   E N N E   A O R B
A P O D S   D E E S   N A S O
```

Puzzle Solution 27

A	S	E	C	■	P	A	R	K	■	T	U	M	M	Y
L	O	G	E	■	L	R	O	N	■	E	I	N	E	N
E	L	A	N	■	A	A	N	I	■	C	R	O	R	E
F	I	L	T	H	Y	W	I	T	C	H	■	P	E	Z
■	■	■	R	A	D	■	N	P	I	N	S	■	■	■
H	A	M	I	T	U	P	■	I	N	O	R	D	E	R
I	S	I	■	H	M	E	E	C	■	T	E	N	O	■
T	A	K	E	A	B	A	C	K	S	E	A	T	T	O
I	D	E	S	■	C	O	E	U	R	■	O	O	N	■
T	A	D	P	O	L	E	■	R	C	A	D	O	M	E
■	■	N	P	E	P	A	■	C	O	O	■	■	■	■
M	A	D	■	P	O	I	N	T	O	F	V	I	E	W
D	R	O	M	O	■	P	A	A	R	■	E	F	G	H
L	O	S	T	S	■	E	I	R	E	■	I	S	E	E
I	D	O	N	E	■	S	L	E	R	■	N	O	R	T

Puzzle Solution 28

H	O	O	F	S	■	T	I	B	I	A	■	M	C	S
A	F	A	T	E	■	U	D	A	L	L	■	I	A	L
B	A	S	K	E	T	B	A	L	L	F	A	N	N	Y
■	■	■	N	T	S	B	■	S	E	R	V	I	C	E
G	E	N	O	V	E	S	E	■	E	A	M	E	S	■
D	L	I	X	■	■	U	S	A	■	T	A	R	T	■
A	B	E	■	M	E	A	C	U	L	P	A	■	■	■
Y	E	R	T	L	E	T	H	E	T	U	R	T	L	E
■	■	E	V	E	A	R	D	E	N	■	R	E	M	■
R	E	M	N	■	E	N	E	■	■	S	A	T	I	■
I	N	E	P	T	■	■	S	N	A	R	L	S	A	T
P	H	O	E	B	U	S	■	O	N	E	A	■	■	■
L	A	W	R	E	N	C	E	S	U	M	M	E	R	S
E	L	E	■	A	I	O	L	I	■	I	O	D	I	C
Y	O	D	■	M	S	T	A	R	■	T	N	U	T	S

Puzzle Solution 29

```
G E N L ■ S A M M ■ B A F T A
M A L A ■ O J A I ■ A P R E S
A T E M ■ N O R N ■ L A O A T
C A R P A T B A G G E R S ■
■ ■ B R A ■ ■ N E T T L E
E T S ■ A G R E E O N ■ N Y M
N A H U M ■ I C E T ■ M I S O
W H E R E S T H E E G G X I T
R I D E ■ C A E N ■ R O O N E
A T S ■ R O S S S E A ■ N E D
P I A G E T ■ ■ N F C ■
■ T H E S O U L O F T W I T
A P E E K ■ O D E R ■ I A M A
L L A N O ■ H O L M ■ E N E O
P A R T F ■ S N Y E ■ S E T S
```

Puzzle Solution 30

```
E N E R ■ A H A T ■ W A K E S
M E S A ■ D O G E ■ H R O S S
T T T T ■ D R A X ■ E A S E S
S W E E P S S T A L L S ■
■ T E D I U ■ H S I A ■ S S H
■ A P P L A U D N O E N D
G T S ■ E T D S ■ ■ I E A T
H E T E R O S ■ H E N R Y I V
O R O F ■ ■ O G L E ■ A L S
S P O T P O T S T O P S ■
T S P ■ N E E T ■ N A N C I
■ G E R A L D G L A R E D
M A C A U ■ S E E A ■ C E R E
B Y G U M ■ E R A T ■ K E E N
A S I D O ■ S S E E ■ S L I T
```

Puzzle Solution 31

S	A	N	D	L		S	S	S	S		M	T	N	S
A	C	O	O	L		A	T	T	A		A	R	E	I
L	O	R	N	A		N	E	U	T		G	A	I	L
O	P	I	U	M	P	O	P	P	Y		A	N	G	E
		T	A	I			Y	R	L	Z	Q	H	X	
E	G	O		S	C	A	R		S	E	I			
D	A	R	A		O	L	E	G		S	N	O	O	T
H	O	L	D	I	N	G	P	A	T	T	E	R	N	S
S	L	Y	A	S		A	L	P	H		S	R	Z	U
			P	M	S		Y	E	E	S		S	E	P
I	N	S	T	Y	L	E			S	H	S			
S	S	E	E		O	M	E	N	C	O	M	I	C	S
L	I	E	D		P	O	P	A		R	A	D	I	O
E	D	I	T		E	N	O	L		T	R	E	E	D
T	E	T	O		S	O	S	A		O	M	E	N	S

Puzzle Solution 32

A	C	E	R		T	A	R	O		S	E	T	O	F
L	A	R	C		A	B	A	S		A	G	O	R	A
T	R	A	V	E	L	S	S	O	S	L	O	W	L	Y
A	N	O	R	A	K		P	N	C		I	S	E	E
R	E	F		U	S	X		G	O	A	D			
			S	H	A	R		P	L	E	A	S	E	
A	R	M		O	N	E	A		L	A	B	O	R	
T	H	E	C	O	W	A	R	D	L	Y	L	I	O	N
E	Y	E	R	S		X	O	R	O			G	T	I
E	S	T	E	E	M		W	A	S	I				
		A	E	A	N		W	T	O		S	H	A	
C	A	T	T		M	E	I		A	L	E	A	S	T
T	R	A	I	N	E	D	A	S	S	A	S	S	I	N
R	E	E	V	E		D	I	K	E		O	H	N	O
S	E	R	E	R		A	M	D	A		S	A	G	S

Puzzle Solution 33

P	I	E	R	■	E	D	A	M	■	Q	U	I	T	S
R	O	M	O	■	S	R	Z	U	■	U	L	N	A	E
O	W	E	S	■	T	A	U	T	■	A	T	A	R	I
M	A	R	I	N	E	G	R	E	E	N	■	B	O	N
■	■	E	Y	E	S	■	A	T	A	S	T	E		
K	N	O	R	R	■	G	Y	R	A	T	E			
A	T	V	■	O	U	T	E	A	T	■	I	N	S	O
T	O	E	D	■	N	A	T	C	H	■	P	T	U	I
T	F	R	S	■	E	R	I	K	A	S	■	I	L	L
■	B	O	G	A	R	T	■	P	E	A	K	S		
J	O	U	S	T	S	■	C	P	A	S				
A	P	R	■	C	E	L	E	R	Y	S	T	A	L	K
H	E	D	D	A	■	A	L	O	G	■	H	S	I	A
A	N	E	A	R	■	T	A	S	M	■	E	Y	E	R
N	A	N	O	S	■	I	S	S	Y	■	R	E	D	O

Puzzle Solution 34

B	E	A	U	T	■	P	E	W	I	T	■	N	E	N
O	R	C	H	S	■	U	N	I	T	E	■	A	S	E
P	L	O	W	E	R	P	A	N	T	S	■	P	A	X
H	E	S	■	L	I	A	■	A	T	E	O	U	T	
A	S	T	A	I	R	E	S	■	K	E	N	L		
■	P	O	E	■	E	M	E	R	G	E	N	T		
N	Y	M	E	T	■	D	A	I	S	■	S	O	A	R
F	O	E	■	T	E	S	L	A	■	N	N	E		
L	G	T	H	■	R	A	I	D	■	P	A	I	G	E
D	I	R	E	N	E	E	D	■	A	O	L			
■	O	E	Y	E	■	E	C	L	I	P	S	E	D	
S	N	A	P	A	T	■	E	O	N	■	T	G	I	
O	E	R	■	L	O	S	T	N	E	T	W	O	R	K
M	A	E	■	A	P	A	R	T	■	A	L	L	E	E
E	L	A	■	S	S	T	A	R	■	T	E	A	T	S

Puzzle Solution 35

Puzzle Solution 36

Puzzle Solution 37

G	A	R	N			I	L	L	U	S		D	P	S
A	L	E	T	A		N	A	S	S	A		R	E	O
L	I	P	S	M	A	C	K	I	N	G		E	A	D
B	A	R	B	E	C	U	E			I	B	A	R	S
A	S	O		C	A	B		T	E	N	A	M		
			C	H	R	I	S	T	M	A	S	B	U	S
D	A	C	H	E		I	T	A	W		O	P	A	
O	L	E	O		S	D	L	T	G		H	A	U	S
L	T	S		A	M	A	T			H	T	T	P	S
L	I	T	T	L	E	B	S	A	M	B	O			
		S	O	L	E	A		M	O	A		M	A	O
I	R	I	D	S		R	O	C	K	S	A	L	T	
L	O	B		T	R	A	D	E	S	E	C	R	E	T
A	B	O		A	C	R	A	B		D	A	N	N	O
W	O	N		R	A	I	S	A			D	I	E	M

Puzzle Solution 38

B	E	A	T		E	G	G	A	R		C	A	W	S
A	T	L	I		A	T	T	W	O		O	D	B	O
T	H	E	G	A	R	D	E	N	O	F	E	D	E	N
E	N	G	R	A	M			S	O	U	N	D	S	
		E	A	U	S		B	T	O	R				
H	A	I		A	F	C		E	S	D		T	Y	S
E	R	N	I		F	A	R	R		I	C	O	M	E
D	E	N	N	I	S	T	H	E	M	E	N	A	C	E
G	W	Y	N	N		M	E	A	N		N	I	A	S
E	E	C		T	E	A		V	E	N		R	S	T
		Y	A	N	N		E	M	T	S				
C	R	O	A	K	S				O	S	T	E	N	D
T	H	E	P	E	L	I	C	A	N	B	R	I	E	F
R	U	N	A		E	C	O	L	I		I	N	S	C
L	E	S	T		R	U	N	I	C		P	E	I	S

Puzzle Solution 39

```
T A M T A M   S K E E   B I D
A N G O R A   T A M S   A T E
C E S S N A   A R P S   T H E
    S E M I D E T A C H E D
H W F E     A T N   M O E S
Y A R D A R M   B E L G
M Y A   N A S H   N A V A J O
A N T O N I O B A N D E R A S
N E S S E S   S N E D   C R O
    A S A S   G A S O H O L
S H A K   S R I   H Y D E
B R E A K T H R O U G H
A A S   E R I S   D I E O F F
C C I   R I F T   O G L A L A
K E R   N O T A   S A L M O N
```

Puzzle Solution 40

```
T R A N   C I T O   T L M A
H E S A   A G A I   H E A P
E A T A   M A L L E A B L E
A L A   E R S K I N E   P R O
B L I M E Y   L E V   B A C H
C O R E L   C I R   I L I U M
S T E M   E I K   G L A S S Y
    O L D G E E Z E R
A U T I S M   A L O   I A M B
S P Y R I   R P I   C N D O S
A R P S   B O I   P A G E R N
S A E   M I S R E A D   L A E
  I D O N T C A R E   C P L S
  S I R E   O T B S   W H E T
  E N A M   E E E E   M I S S
```

Puzzle Solution 41

```
B D R M ■ Z A G S ■ S B A C K
V I C I ■ ■ O P R Y ■ E H L U A
D E V E L O P E R ■ ■ L O V I N
S U R N A M E S ■ E L U A N T
■ ■ M E L ■ ■ S S S S ■
M I F F E D ■ M U S ■ E S T R
C P L U S ■ K A N A ■ ■ H O E
K O O L A I D D R I N K E R S
E D A ■ N C A A ■ O E I R O
E S T O ■ E F T ■ H B A K E D
■ ■ V E R E ■ O R U ■
E O C E N E ■ B A R T E R E D
F T O R D ■ U P T O S N U F F
G E N I I ■ H O E R ■ E E O C
H A F T S ■ W E S S ■ S S R S
```

Puzzle Solution 42

```
I G A S ■ I A S I ■ S I X A M
M E T A ■ F R O M ■ E N R O N
E A R N ■ S T R A ■ G O A L O
T R I C K O R T R E A T Y ■
A B A T E ■ ■ I L L ■ S T D
■ ■ I N S T S ■ I S O P O D
A C A ■ T T O P S ■ M E N D
W O R D S P R O N O U N C E D
A P B S ■ E C A R D ■ S Y D
S T I C K Y ■ K G R I P ■
H O T ■ E E S ■ ■ N L A K E
■ R O A S T E D P E A N U T
T E A O R ■ A L I E ■ N E B R
U P T O N ■ N Y E T ■ E L L E
T H E O S ■ E S M E ■ S E A S
```

Puzzle Solution 43

M	E	A	G		A	T	T	I	C		U	S	P	S
E	R	D	E		F	R	E	R	E		M	K	N	G
C	R	I	T		K	A	R	O	L		B	E	E	T
C	O	M	M	E	R	C	I	A	L	C	O	W		
A	R	E	A	L			M	O	E		E	L	M	
		D	I	S	C	S		E	I	R	E	A		
S	A	L		A	N	Y	Q	U	E	S	T	I	O	N
I	D	O	L		A	T	U	N	E		I	N	R	O
D	I	N	I	N	G	O	A	T	E	S		G	A	R
L	E	G	T	O		W	O	E	I	S				
E	T	S		A	I	M		S	C	A	B	S		
	P	U	M	P	K	I	N	P	I	R	A	T	E	
P	I	O	N		A	U	R	A	L		I	M	E	T
E	V	O	E		S	O	M	N	I		M	C	A	T
S	S	N	S		S	G	A	T	E		P	O	M	S

Puzzle Solution 44

A	B	B	A		M	M	D	C		M	R	F	O	X
T	E	R	R		A	I	R	H		A	H	O	R	A
T	A	E	M		S	T	O	A		H	E	X	E	S
I	N	V	O	I	C	E	O	F	D	O	O	M		
C	O	E	U	R		L	E	O	G		U	S	B	
	R	E	G	D		S	E	A	B	L	U	E		
S	S	S		M	O	L		N	I	D	R	E		
N	A	M	E	I	C	A	L	L	M	Y	S	E	L	F
A	B	A	R	N		B	I	B		R	Y	S		
K	O	S	O	V	A	R		B	E	L	K			
E	T	H		A	C	E	S		B	I	D	E	S	
	H	A	L	T	E	R	E	D	S	T	A	T	E	
C	H	I	L	I		K	T	C	O		C	R	U	D
F	E	T	I	D		E	A	T	S		A	I	D	E
L	Y	S	E	S		D	S	O	S		R	A	E	R

Puzzle Solution 45

R	A	N	I	D		B	A	C	K		P	E	R	L
O	F	U	S	E		O	L	L	A		O	B	O	E
B	L	I	N	D	A	L	L	E	Y		C	O	N	G
T	E	T	T	E	T	E		O	S	M	O	N	D	S
			E	T	Y	M		E	N	D	E	D		
N	S	E	W		U	N	D	E	R	D	O			
E	P	A	C	T		C	L	I	I		A	N	G	
S	U	R	V	E	I	L	L	A	N	C	E	A	I	D
I	R	S		N	M	E	X		I	V	A	N	A	
	O	N	E	O	V	E	R		E	S	S	Y		
A	B	A	C	I		I	D	O	S					
R	E	S	T	S	O	N		E	B	W	H	I	T	E
I	G	L	U		T	O	P	S	B	O	T	T	O	M
D	U	A	L		I	R	A	S		R	E	S	I	T
E	N	N	E		C	W	L	A		K	N	O	T	S

Puzzle Solution 46

D	C	X	L		P	I	U	S		M	C	R	I	B
O	S	E	E		A	R	O	O		I	F	E	L	L
N	H	R	A		G	A	F	F		D	O	N	O	T
T	A	X	F	R	E	E	A	T	L	A	S	T		
G	R	E	Y	S			G	A	I		A	L	S	
O	P	S		I	B	M	S		A	R	A	B	I	C
		N	D	H	J	E	P		A	L	E	A		
M	A	R	K	E	T	G	A	R	D	E	N	E	R	S
O	L	E	G		R	I	O	T	E	D				
Y	E	S	B	U	T		R	A	S	E		O	E	O
A	X	E		N	I	M			N	O	M	A	N	
		M	A	D	C	O	W	D	I	S	E	A	S	E
K	R	B	P	E		P	I	O	N		U	N	I	S
M	A	L	I	A		E	F	A	X		F	I	N	E
S	P	E	E	D		S	I	S	S		S	S	G	T

Puzzle Solution 47

```
C O W L   I M M A D   S T B D
A T O E   T E N C C   A R O O
P R E S B Y T E R I A N I S M
O A T E R   E M I   T A S S E
T S O   A D D E D U P   H A D
    L G E       S A E
  D I E A N O T H E R D A Y
G E N O   E L I A N   B M O C
U L T   A B A N D O N   A G S
A T E A R   V E A   G E N I I
R A R I N G       R E P A S S
  B R O N Z E M E D A L
S L A W   A A R O N   C O B S
R A N A   T C O A T   T N T S
O A K Y   S H O N E   S E W N
```

Puzzle Solution 48

```
I S S H E   D O A S   T U F T
R A T O N   A L L A   O S L O
O B O L S   R A U L   S T E R
V O W E L L E N G T H   E X T
    E R I     I A G R E E
V E L C R O   P O N I A R D S
E C U A   N I P P E R S
A K I T A   A M I   Y M C A S
    S T A L K E D   E E R O
C A P I T A L Q   C A N O E D
I N S T I R   V C R
N A A   C C H A S C C H C A N
E N L S   T E R I   H O A G Y
M A M E   I B E G   E N D U P
A S S N   C R I N   D E S E D
```

Puzzle Solution 49

D	I	S	C		O	P	E	D	S		N	A	O	H
R	E	M	O		N	O	R	A	H		A	N	T	E
I	R	A	N		L	O	N	G	E	S	T	D	A	Y
B	I	S	C	A	Y	N	E		Q	U	A	Y	S	
		E	R	I	A		A	Z	U	R				
A	R	F	A	R	F		A	C	T	A	E	O	N	
G	A	O	L	S		A	B	E	E	T		R	E	Y
A	N	N	S		A	L	A	N	W		S	E	A	R
L	C	D		O	V	E	T	T		A	T	O	L	L
	H	A	Y	R	I	D	E		O	C	A	S	E	Y
		S	O	S	A		F	M	A	N				
S	P	A	H	N		U	R	E	T	H	A	N	E	
K	A	R	A	O	K	E	B	A	R		O	L	I	N
I	S	U	P		L	I	E	U	T		P	A	C	E
M	T	G	E		U	S	R	D	A		E	E	K	S

Puzzle Solution 50

S	N	C	C		M	A	S	A	I		I	T	I	N
O	U	L	U		E	N	C	E	S		A	R	S	E
S	M	E	E		S	C	H	O	O	L	M	A	T	E
A	B	O	R	C		A	L	L	E		C	I	D	
		V	E	N	I	V	I	D	I	V	I	C	I	
A	L	T	O	S	A	X		S	E	S	E			
L	E	A		A	R	I	D			L	E	C	H	
L	O	W	E	R	C	A	N	A	D	A	D	A	D	A
S	S	S	S			Y	U	L	E		S	L	E	
	A	G	T	S		E	I	D	E	T	I	C		
K	I	R	S	T	I	E	A	L	L	E	Y			
E	L	E		E	E	R	S			S	E	E	R	S
N	E	C	T	A	R	I	N	E	S		L	I	F	T
A	F	T	A		O	N	E	R	A		E	N	D	E
I	T	O	O		D	E	W	E	D		T	E	S	S